Lectures on Mathematics

W9-CWV-351

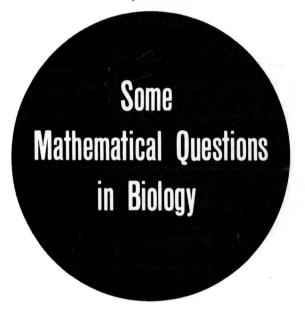

Some Mathematical Questions in Biology

in the Life Sciences
Volume 14

Some Mathematical Questions
in Biology

Lectures on Mathematics in the Life Sciences

Volume 14

Some Mathematical Questions in Biology

The American Mathematical Society
Providence, Rhode Island

Proceedings of the 1981 Symposium on
Mathematical Biology held at the Annual Meeting of the
American Association for the Advancement of Science
in Toronto, Canada, January 8, 1981.

edited by
Stephen Childress

Library of Congress Catalog Card Number 77-25086
International Standard Book Number 0-8218-1164-9
International Standard Serial Number 0075-8523
1980 Mathematics Subject Classifications: 92–06, 93A05, 92A09.

To the memory of Sol I. Rubinow.

CONTENTS

FOREWORD

This volume contains the lectures which were presented at the Fifteenth Symposium on Some Mathematical Questions in Biology, held on January 8, 1981 in Toronto, Canada, in conjunction with the annual meeting of the American Association for the Advancement of Science. The symposium was jointly organized and sponsored by the American Mathematical Society and the Society for Industrial and Applied Mathematics under the auspices of Section A, Mathematics, of the AAAS.

The papers presented in this Symposium deal with two principal areas of theoretical biology, the first focusing on several problems of developmental biology, the second on recent work in biomechanics. Meinhardt and Gierer consider the organization of pattern, using models extending their celebrated work on regulation in systems of reaction-diffusion equations. Emphasis in their presentation is on the role of sequencing as a means of generating structures with a high degree of internal regulation. Lacker and Peskin treat a related but unusual example of modeling, whose aim is to explain the process of control of ovulation number (number of maturing follicles per cycle) in mammals. A system of differential equations is used to describe the regulation of follicle growth by circulating hormones. In a sense Lacker and Peskin also are concerned with (temporal) sequencing of pattern as determined by the integral number of maturing follicles. The third paper in this set, by Percus and myself, deals with some elementary mechanical models of the movement of cells and cell aggregates. Here the interest is in exploring the morphogenetic consequences of relatively simple instructions at the cell level, involving measures of cell-cell adhesion.

The lectures by Koehl and Weinbaum deal with quite different

problems of biomechanics, but there is common ground in their innova-
tive application of classical properties of viscous fluids moving at low
Reynolds numbers. Koehl is interested in the feeding behaviors of
copepods and their relation to food particle size. The complexity of
behavior associated with particle capture poses a number of intriguing
questions for theoreticians interested in time-dependent Stokes flows,
which raises issues complementary to theories of propulsion at low
Reynolds number. Weinbaum discusses methods which are well suited
for such applications, and gives a comprehensive review of their implica-
tions for pore-particle problems arising in models of transport through
membranes.

The final lecture, by Mochon, deals with the mechanics of human
locomotion in a series of models of increasing complexity, and highlights
the useful role which this kind of analysis can play in many related
problems of human movement.

On behalf of all the participants I would like to extend our thanks
to the staff of the American Mathematical Society for their capable
assistance with all phases of planning the Symposium and preparing
these Proceedings.

Stephen Childress

Courant Institute of Mathematical Sciences
New York University
June 18, 1981

Lectures on Mathematics in the Life Sciences
Volume 14, 1981

GENERATION OF SPATIAL SEQUENCES OF STRUCTURES
DURING DEVELOPMENT OF HIGHER ORGANISMS

Hans Meinhardt and Alfred Gierer

ABSTRACT. Sequences of structures are frequently encountered during
development of higher organisms. Different mechanisms can be envis-
aged for its formation: (i) the two-step process of generation of
positional information and its interpretation and (ii) the direct
sequential induction of structures. Positional information can be
generated by local autocatalysis combined with long range inhibition or by cooperation of compartments. Mutual induction of structures requires a long range activation of locally exclusive states.
This mechanism seems to be appropriate for developmental systems in
which experimental evidence indicates that a control of a correct
neighbourhood of structures exists. It will be shown under which
conditions a sequence of structures is more stable than one very
large element or an alternation of only two structures. This mechanism allows a self-regulatory intercalary regeneration of missing
elements. In a two-dimensional field of cells, a stable stripe-
like arrangement of structures with a long extension in one dimension and a short extension in the other can be obtained. The proposed interactions are provided in the form of coupled non-linear
partial differential equations. Computer simulations show that the
proposed models are free of internal contradictions and that their
regulatory behavior is in agreement with experimental observations.

1. POSITIONAL INFORMATION VERSUS DIRECT INDUCTION OF ADJACENT STRUCTURES.

The complexity of structures of a higher organism must be generated

during its development; it cannot already exist in a hidden form within

the egg cell. The generation of structures signifies that at a particular

location something is formed which is not formed in the surrounding. Ex-

perimental evidence suggests that whole sequences of structures are under

a common developmental control. For instance, the implantation of a

second "organizer" (Spemann, 1938) consisting of a small piece of tissue

derived from the dorsal lip of an amphibian blastopore, can induce a com-

plete new embryo. A common developmental control for many structures

which are arranged in an ordered sequence appears necessary not only to

form every structure of an organism but also to obtain their correct spa-

tial relationship. A mechanism which is able to generate sequences of

structures in space consists in the formation of a graded distribution of

a substance, the morphogen, across a field of cells. The local concentration can provide positional information and determine the future pathway
of the cells (Wolpert, 1969, 1971). Many experiments concerning early
insect embryogenesis are quantitatively explicable under this assumption
(Meinhardt, 1977).

A characteristic property of a positional information scheme is that
cell determination is affected only by the local concentration of the
morphogen but not by the state of determination of cells in the neighbourhood. The corresponding behavior has been observed in many instances;
for instance, upon ligation of early insect embryos, regeneration of the
sequence of segments is incomplete, leaving gaps which are not repaired
(Sander, 1975). Similarly, gaps occurring spontaneously in "bicaudal"-
embryos of Drosophila (Nüsslein-Volhard, 1977) remain unrepaired.

Other developmental systems behave differently. For instance, portions
removed from a particular segment of an insect leg regenerate. Confronta-
tion of normally non-adjacent cells initiates a respecification (and frequently a proliferation) of the cells at the discontinuity which leads to
a repair of the gap.

We have proposed molecular mechanisms for the generation and interpretation of positional information and for the control of neighbouring
structure (Gierer and Meinhardt, 1972, 1974; Meinhardt and Gierer, 1974,
1980; Meinhardt, 1977, 1978a,b). In the present article, we provide a
review of these models, pointing out differences and similarities and
compare the behavior of these models with some key experiments.

2. PATTERN FORMATION BY AUTOCATALYSIS AND LATERAL INHIBITION. We have
shown (Gierer and Meinhardt, 1972, 1974) that patterned distributions of
substances can be generated by an interaction of a short-ranging autocatalytic and a long-ranging antagonistic substance. Some possible interactions should be discussed briefly to demonstrate similarities and
differences with mechanisms enabling control of adjacent structures.

The following interaction between an autocatalytic activator $a(x)$ and
the inhibitor $h(x)$ leads to stable patterns:

$$\frac{\partial a}{\partial t} = \frac{ca^2}{h} - \mu a + D_a \frac{\partial a^2}{\partial x^2} \tag{1a}$$

$$\frac{\partial h}{\partial t} = ca^2 - \nu h + D_h \frac{\partial^2 h}{\partial x^2} \tag{1b}$$

The reaction is such that in a small field (smoothing out of any spatial inhomogeneity by diffusion) only a single stable steady state is possible corresponding to a near-uniform distribution in space. In a larger field, the local autocatalysis allows an amplification of small deviations from the steady state while due to the long-ranging inhibition, an activator increase in one part of the field is coupled to a decrease in other parts until the pattern becomes stabilized. It is a property of such a system that in a growing field, after surpassing of a critical extension, an activator maximum is formed at one margin, leading to a graded - and thus polar - concentration profile (Fig.1a). An activator maximum has many properties of the classical organizer: it can "regenerate" after its removal. The regeneration of hydra tissue, and the induction of secondary axis in hydra, show properties characteristic for this type of pattern formation. Another property of a classical organizer is the possibility

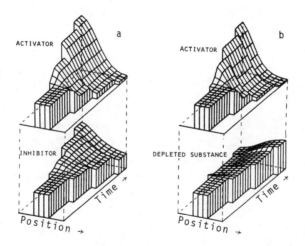

Fig.1: Pattern formation by autocatalysis and lateral inhibition. The interaction between the autocatalytic activator and either the inhibitor (Fig.a, Eq.1) or the depleted substance (Fig.b, Eq.2) leads to a stable patterned distribution of both substances. Assumed is a linear array of cells growing at both margins; the concentrations are plotted as function of position and time. Pattern formation requires an area exceeding that of the activator range. At that size, the activator maximum appears at one boundary of the field since a non-marginal maximum would require space for two slopes. Therefore, this mechanism can generate positional information changing monotonically in the field of cells to be organized.

of its induction by unspecific manipulations. According to the model, a
local inhibitor decrease caused, for instance, by injury or local irradi-
ation can lead to a second activator maximum (second organizer region).

The effect antagonistic to autocatalysis need not be provided by an
inhibitor but may result from the depletion of a precursor s required for
the synthesis of the activator:

$$\frac{\partial a}{\partial t} = ca^2 s - \mu a + D_a \frac{\partial^2 a}{\partial x^2} \qquad (2a)$$

$$\frac{\partial s}{\partial t} = c_o - ca^2 s - \nu s + D_s \frac{\partial^2 s}{\partial x^2} \qquad (2b)$$

A simulation is shown in Fig.1b. This reaction has somewhat different
properties, sometimes distinguishable on the basis of experimental observations (Meinhardt, 1978a). For instance, additional maxima cannot be
induced by an unspecific reduction caused by radiation damage or leakage
of one of the components.

The autocatalytic effect need not be due to a single component, but
may be hidden in a reaction chain. An example is given by two substances,
a and b, each repressing the production of the other. An increase of
a leads to a decrease in production of b and the lowering of b
leads to a further increase of the production of a. Pattern formation
requires that an increase of a at a particular location is coupled to a
decrease of a in the surroundings, requiring a further substance with a
long diffusion range. An example of an interaction scheme of this type in
which no component is autocatalytic per se is the following:

$$\frac{\partial a}{\partial t} = \frac{\mu}{\kappa + b^2} - \mu a \qquad (3a)$$

$$\frac{\partial b}{\partial t} = \frac{\mu \cdot c}{\kappa + a} - \mu b \qquad (3b)$$

$$\frac{\partial c}{\partial t} = \nu(a - c) + D_c \frac{\partial^2 c}{\partial x^2} \qquad (3c)$$

The substance c supports the production of b but in the system as a whole
it has the function of a long-range inhibitor since it is antagonistic to
the (hidden) autocatalysis. Fig.2 shows the formation of a regular periodic structure in a growing field according to this interaction.

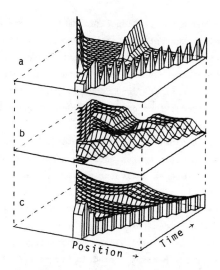

Fig.2: Formation of a periodic pattern in a growing array of cells.
Two substances are assumed (plotted in the corresponding subpictures
a and b) which inhibit each others production (Eq.3). This inhibition
of an inhibition is equivalent to an autocatalytic reaction. A long-
ranging substance (Fig.c) stabilizes the pattern since it is antagonistic to this hidden autocatalysis. Starting from an almost homogene-
ous distributions, with growth, the system passes through a polar and
a symmetrical pattern until regular periodic patterns occur.

3. POSITIONAL INFORMATION AND ITS INTERPRETATION. As mentioned, pattern
formation by autocatalysis and lateral inhibition allows the formation of
a graded distribution of a substance. The local concentration is a
measure for the distance from the maximum and can therefore be used as
positional information: If the selection of a particular developmental
pathway is under the control of the local concentration of such a morphogen, spatial sequences of structures can be specified.

A challenging biological system to test such a model is the determination of segments in a developing insect embryo. Experiments have been
performed in which the normal development has been disturbed (see Sander,
1976) leading to an abberant development. It was shown that many of these
experiments become quantitatively explainable under the assumption that
the activator maximum is confined to a small fraction of the egg located
at the posterior (rear) pole and that the long-ranging inhibitor with its
graded distribution across the whole egg supplies the proper positional
information (Meinhardt, 1977). The explanatory power of that model is

illustrated with one example. Unspecific manipulations such as UV-irradiation, puncturing of the anterior pole or centrifugation of the egg can
lead to a completely symmetrical development in which the head and the
thoracic segments become replaced by a second, mirror-symmetric set of
abdominal structures (Kalthoff and Sander, 1968; Schmidt et al., 1975;
Rau and Kalthoff, 1980). According to the model, the activator maximum is
located at the posterior (abdominal) pole; therefore, the inhibitor concentration has its minimal value at the anterior pole. There, any further
reduction of the inhibitor concentration can trigger autocatalytic activation, leading to a second activator maximum which causes a second ab-
domen to be formed. In agreement with the model, the induction of the
second maximum is an all-or-none effect and the pattern of segments
formed is nearly independent of the dose of UV-irradiation. The transition from a polar to a symmetrical pattern is connected with a characteristic alteration of the segment formed in the center of the egg. This
alteration is described in a straightforward manner by a positional information scheme. Let us enumerate the segments from the head to the ab-
domen 1,2...16. The segment formed in the center of a normally developing
egg is the segment 5, a thoracic segment. After induction of a symmetrical development, the plain of symmetry is around segment 9, leading to a
pattern 16...9...16. According to the model, after induction of the
second activator maximum, inhibitor diffuses from both sites into the
center, leading to an increase of the inhibitor concentration (positional
information) and therefore to the formation of more posterior structures
there.

The final spatial complexity of an organism is not generated by the
interpretation of just one gradient or of two orthogonal gradients. Additional positional information systems are required to determine the
finer details. The interpretation of the primary positional information
leads to a subdivision into discrete patches of differently determined
cells, and positional information for subpatterns can be generated, within the patches, by a cooperation of such "compartments" (Meinhardt,
1980). The common boundary region of at least two such patches may become
a new source region of further morphogens. Many experiments concerning
the determination of insect appendages are explainable under this assumption.

4. GENERATION OF SEQUENCES OF STRUCTURES BY LONG-RANGE ACTIVATION OF
LOCALLY EXCLUSIVE STATES. While in a positional information scheme the
cells ignore the determination of their neighbours and respond only to a
local morphogen concentration, some developmental systems indicate a
strong control of correct neighbourhood. An example is the proximo-distal
as well as the circumferential sequence of structures within an insect
leg. Artificially induced gaps are repaired (Bohn, 1965; French, 1978).
Moreover, gaps produced by grafting of surplus structures on an existing
stump lead to a duplication of the excessive structures (Fig.3). To give
a brief description of some experimental key results let us denote the
sequence of structures such as parts of a leg segment, in proximo (close
to the body)-distal dimension, with 123...9. An experimentally produced
leg segment containing the structures 12/89 will replace the missing
structures by intercalary regeneration, forming the complete sequence
123456789 (the intercalated elements are underlined). The corresponding
experiment with surplus structures, e.g. 1234567/23456789 would lead to
an even longer leg 1234567654323456789. Removal of an internal part of

a leg will never occur under natural circumstances such as the bite of a
predator. That the organism is nevertheless able to regenerate suggest
that the role of this mechamism is not the repair after artificial remo-

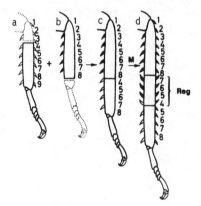

Fig.3: Duplication of excessive parts of a cockroach leg (after Bohn,
1970). Experimental juxtaposition of originally non-adjacent pattern
elements (a-c) leads after one or two molts (M) to intercalary regeneration of the missing elements (Reg) at the site of the gap (d). This
experimental result indicates that it is not the normal pattern of the
leg segment but the neighbourhood of elements of the sequence of
structures which is under developmental control.

val of parts but that it is a component in the generation of the structures during normal, undisturbed embryonic development. For instance, it
is possible that initially only the terminal structures, e.g. 1 and 9 are
layed down and that the remaining structures are filled in by intercalation.

The formation and stability of an abnormal mirror-symmetric sequences
such as ... 45676543234 ... is particularly instructive in discriminat-
ing between types of models. For instance, it can be ruled out that the
internal organisation of the leg segment is controlled by a morphogen
gradient, formed by a source at one side and by a sink at the other side
since such a model cannot account for the intercalation of structures in
a reverted sequence. It rather appears that regeneration results from
direct mutual or consecutive induction of neighbouring states. We have
proposed explicit models for this type of pattern formation (Meinhardt
and Gierer 1980). Molecular interactions are introduced such that two or
more locally exclusive states of activation can be generated, with neighbouring states stabilizing each other in analogy to a symbiosis.

5. HOW TO GENERATE STABLE LOCALLY EXCLUSIVE STATES. Cell determination is
considered to be an all-or-none event shifting the cell into a new stable
state that can be transferred to daughter cells upon proliferation. Presumably, determination corresponds to the activation of particular genes
out of a set of alternative genes, the others becoming repressed. The
stability of a state of determination requires some positive feedback of
a gene on its own activity. Alternative feedback loops are assumed to
compete with each other in such a way that only one loop can be activated
in a given cell (Meinhardt, 1978b). A kinetic interaction with this property is described in Eq.4.

$$\frac{dg_i}{dt} = \frac{c_i g_i^2}{r} - \alpha g_i \qquad (4a)$$

$$\frac{dr}{dt} = \sum c_i g_i^2 - \beta r \qquad (4b)$$

Each one of a set of alternative genes i (i = 1,2...n). is activated
by an activator g_i acting autocatalytically on its own production as
described by the term $(c_i g_i^2)$. Autocatalysis is counteracted by the
action of a common repressor r. Each active gene produces and reacts upon
this repressor. It is further assumed that gene activator and repressor

are destroyed by normal first-order kinetics. These equations are formally related to the equations describing the pattern forming reaction (Eq.1). This similarity is not accidental: in pattern formation, the synthesis of a substance is enhanced at a particular location and suppressed at others. In cell determination a particular gene is assumed to be activated and the alternatives repressed, corresponding to pattern formation in an abstract "gene space".

A simple example for two mutually exclusive states is given by the following equation:

$$\frac{\partial g_1}{\partial t} = \frac{\alpha}{\kappa + g_2^2} - \alpha g_1 \qquad\qquad (5a)$$

$$\frac{\partial g_2}{\partial t} = \frac{\alpha}{\kappa + g_1^2} - \alpha g_2 \qquad\qquad (5b)$$

For interactions according to Eq.4 or 5, a state with uniform equal activation of all genes is unstable. Any slight advantage of one state leads to the exclusive activation of this particular state, and to the repression of the alternative state(s).

6. LONG RANGE ACTIVATION OF THE LOCALLY EXCLUSIVE STATES. Such mutual exclusive activation of cell states can lead to defined spatial sequences of different structures if, in addition, cells in any given state cause and stabilize activation of a different state in the neighbourhood via diffusable substances s_i.

$$\frac{\partial g_i}{\partial t} = \frac{c s_j g_i^2}{r} - \alpha g_i + D_g \frac{\partial^2 g}{\partial x^2} \qquad\qquad (6a)$$

$$\frac{\partial s_i}{\partial t} = \gamma g_i - \gamma s_i + D_s \frac{\partial^2 s}{\partial x^2} \qquad\qquad (6b)$$

$$\frac{\partial r}{\partial t} = \sum c s_j g_i - \beta r \quad \text{with } i,j = 1,2;\ i \neq j \qquad\qquad (6c)$$

The (small) diffusion of the g_i-molecules is of importance for size-regulation and for the coherence of g_i-regions.

Let us consider the simplest case of two states (g_1, g_2) and s with a near-uniform steady state with equal concentrations g_1 and g_2

throughout the field. Since s_1 or s_2 distributes over larger space by diffusion, a local increase of g_1 will increase further by self-enhancement, but the accompanied increase in s_1 concentration favours g_2-activation in the surroundings. Therefore, two neighbouring patches of high g_1 and of high g_2 concentration respectively, are formed (Fig.4).

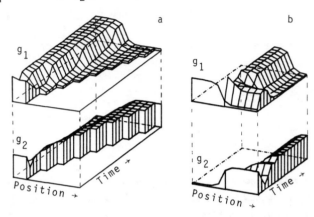

Fig.4: Size-regulation. A system of two feedback loops (g_1 above, g_2 below) which compete with each other locally but support each other over a longrange (Eq.6) show pattern formation with good size regulation. (a) Proliferation of the cells with high g_2-concentration leads to an increased support of the g_1-feedback loop. At the zone of transition between high g_1 and high g_2, some of the cells switch from a high g_2 concentration to a high g_1 concentration and the correct proportion of both areas is maintained. (b) Removal of the cells in which the g_1-loop is active leads to a new pattern formation in the remaining cells, resulting in a new equal partition into a g_1 and a g_2 area.

7. EQUIVALENCE OF MULTICOMPONENT SYSTEMS WITH AUTOCATALYSIS-LATERAL INHIBITION. Pattern formation by mutual lateral activation of locally exclusive states may seem, at first sight, as a mechanism qualitatively different from pattern formation by autocatalysis and lateral inhibition which can lead to gradients and other morphogenetic fields specifying positional information as described in section (2). However, a closer analysis shows that this is not the case; rather, mutually exclusive lateral activation can be subsumed under a generalization of the theory of autocatalysis and lateral inhibition (Gierer, 1981). It was shown by mathematical analysis that the conditions of autocatalysis and lateral inhibition are necessary requirements for the simplest case of two com-

ponents independent of details of models. The conditions can be generalized to some extent to multiple-component systems. The generalization
cannot be based on determining whether or not an individual substance is
autocatalytic. As shown in section (2), patterns can be formed even if no
single component of the system is in itself autocatalytic, because autocatalysis may result from inhibition of inhibition. However, if in a
multicomponent system there are two subsets of components distinguished
by ranges (e.g. due to different diffusion rates) and if the short-range
subset is autocatalytic as a system (perhaps by inhibition of inhibition), whereas the long-range subsystem is crossinhibiting (thus prevent-
ing an overall autocatalytic explosion), spatial patterns can be generated. Requirements are that redistribution of molecules belonging to the
inhibitory subset is sufficiently large and that of the activating subset
sufficiently small. This is a generalization of the two-factor theory of
autocatalysis and lateral inhibition to more than two components. In
molecular terms, it implies that activation and inhibition need not be
properties of individual components but can be features of systems of
such components. The models for sequences of structures based on lateral
activation as described in this paper are of this type.

   This can be directly demonstrated for the case of activation of two
locally exclusive states as exemplified by Eq.(6). Let us introduce a
parameter transformation, with the sums and differences as new parameters: $a_s = g_1+g_2$, $a_d = g_1-g_2$, $s_s = s_1+s_2$, $s_d = s_1-s_2$.
The subsystem a_s, s_s is of no particular interest. The system of the
differences a_d, s_d, however, is closely related to the activator/in-
hibitor system in the context of the theory of lateral inhibition. The
near-uniform state corresponds to $a_d = 0$, $s_d = 0$; upon pattern formation, activating g_1 in one and g_2 in another part of the field, the
difference $a_d = g_1-g_2$ is autocatalytic, forming a __gradient__ across
the field. At the same time, $s_d = s_1-s_2$ forms a gradient of the
same orientation but with a wider range across the field due to larger
diffusion rates. Since s_2 cross-activates g_1 and as s_1 cross-activates g_2, the difference $s_d = s_1-s_2$ __inhibits__ $a_d = g_1-g_2$:
it acts by lateral inhibition. Aside from this qualitative demonstration
of the inhibiting function of s_d, it has been shown that in the linear
approximation of deviations from the uniform distribution there is a
definitive mathematical correspondence of the difference parameter a_d
with activation, and of s_d with inhibition. The mathematical decision

whether patterns are formed is based on this linear approximation. It
follows that, in the context of pattern formation starting from near-
uniform distribution, mutually exclusive lateral activation is mathematically equivalent to lateral inhibition.

8. SIZE REGULATION. The size of the g_1 and g_2 patches are regulated
in relation to the total size of the available space (Fig.4). Imagine
that g_1 is too large and g_2 too small. Then g_2 would get, via the
high s_1, a much larger support compared with the low s_2 concentration
supporting g_1. This has the consequence that some cells reduce g_1
production in favor of g_2 production until the correct relation between
the two is obtained. An essential condition for this size regulation is a
low but finite diffusion of the g_i-molecules. Due to such diffusion,
the transition between the area of high g_1 and high g_2 concentration
is not a step-function; a zone of transition exists between both areas.
This slight "infection" extending from activated areas supplies a low
level in non-activated areas g_i from which autocatalysis can start if
sufficient support by the s_{i+1}-molecules is available.

9. FORMATION OF STRIPES. In a two-dimensional field a reaction according
eq.6 is able to generate stable stripes. This is because a stable pattern requires long common boundaries between the areas of high g_1 and
high g_2 concentration to allow for mutual support by s_2 and s_1
molecules. Patches are thus formed with a long extension in one dimension
and a small extension perpendicular to the first. The width of the
stripes is determined by the range of the mutual support, whereas its
length depends merely on the extension of the field. Small spatial cues
can determine localization and orientation of the stripes. Fig.5 shows
the stripes initiated by small random fluctuations, or by a local stimu-
lus of g_1. In the latter case, alternating concentric rings of high g_1
or high g_2 concentrations are formed.

10. FORMATION OF A GRADIENT ACROSS THE SMALL EXTENSION OF A FIELD. Inti-
mately connected with the ability to form stripes is the ability to generate a gradient oriented along the short extension of a field. The problem will be exemplified by regarding the geometry of an insect egg. The
ratio of width to length is about 1 : 3. Organization of the antero-
posterior dimension (long axis) can be due to a simple monotonic gradient

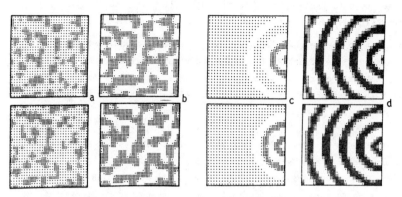

Fig.5: Formation of "stripes". The long-range activation of locally exclusive states favors the formation of stripes since the mutual activation requires long common boundary regions. Density of dots indicates concentrations of g_1, or g_2, respectively. (a,b) Complementary pattern (g_1 above, g_2 below) in a two-dimensional array of cells, initiated by random fluctuations. (a) intermediate state (b) final stable state. (c,d) similar as (a,b) but initiated by a small local g_1 increase at the right side of the field. (After Meinhardt and Gierer, 1980)

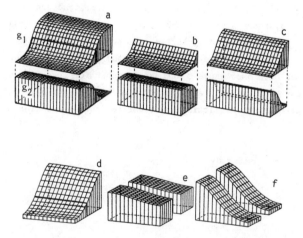

Fig.6: Gradient formation by lateral activation (Fig.a-c, Eq.6) and by lateral inhibition (Fig.d-f, Eq.1) can behave differently in a two-dimensional field of cells. A quadratic field is bisected (heavy lines in a and d). In the lateral activation scheme, a graded distribution of g_1 and g_2 is formed along the short dimension of the field. In the lateral inhibition mechanism (d-f), the remaining activator pattern degenerates along the short dimension and a new graded distribution is formed, upon random initiation, along the long extension of the field.

formed by autocatalysis and lateral inhibition along the longest dimension (Fig.6). On the other hand, such mechanism cannot lead to a stable

gradient in the narrow (dorso-ventral) dimension of an isotropic field
because a ridge of activation would decay into several peaks. However,
the mechanism of mutual activation of locally exclusive states is able to
generate a pattern in the short-(dorso-ventral) dimension as shown in
Fig.6.

11. SEQUENCES OF STRUCTURES. A sequence of structures can be described as
an array of cells in which, in a terminal group of cells, a gene no.1 is
active, in neighbouring cells gene no.2, followed by cells in which gene
no.3 is active and so on. Generating such a structure requires that the
sequence 123... is more stable than the situation in which the same gene
is active in all cells, or an alternation of only two active states e.g.
1212... occurs in space. Further the correct sequence 123... must be more
stable than an arrangement in which every structure is present but with
an incorrect neighbourhood, e.g. 1324... The condition of mutual support
by molecules s_i (Eq.7) enforces the neighbourhood, for structure i, of
either i+1 or i-1. To avoid the alternation of two structures we have to
introduce an asymmetry such that each gene has a stronger tendency to
activate the gene corresponding to the structure following in the sequence compared with those corresponding to the preceding element. For
instance, gene no.2 must cause a stronger support for gene no.3 compared
with those for gene no.1. An example for a model generating defined sequences is given by the following equations (Meinhardt and Gierer, 1980).

$$\frac{\partial g_i}{\partial t} = \frac{c_i {g_i'}^2}{r} - \alpha g_i + D_{g_i}\frac{\partial^2 g_i}{\partial x^2} \qquad (7a)$$

$$\text{with } g_i' = g_i + \delta^- s_{i-1} + \delta^+ s_{i+1}$$

$$\frac{\partial s_i}{\partial t} = \gamma(g_i - s_i) + D_s\frac{\partial^2 s_i}{\partial x^2} \qquad (7b)$$

$$\frac{\partial r}{\partial t} = \sum c_i {g_i'}^2 - \beta r \qquad (7c)$$

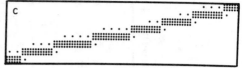

Position →

Fig.7: An ordered sequences of active feedback loops ("genes") can be activated in a growing array of cells. The concentration of a particular gene activator g_i (Eq.7) is indicated by the density of dots. Whenever sufficient cells are present in which a particular loop is active, the lateral activation becomes sufficient for the induction of the following loop.

The stronger support of the next following element of the sequence requires that $\delta^- > \delta^+$. The generation of a sequence of structures resulting from this type of interaction is shown in Fig.7.

12. CONDITION FOR INTERCALARY REGENERATION. The biological example given in Fig.3 shows not only that a sequence of structure is formed during development but that parts removed by excision are replaced. The introduced asymmetry $\delta^- > \delta^+$ is not yet sufficient to explain this repair. Imagine a sequence 123/789. At the gap, no.3 tends to induce no.4 and no.7 tends to induce 8. A repair however, would require that only structure 4 but not 8 is formed. This occurs if lower states have an advantage in competition with higher states. The advantage can be modelled for by introducing, into eq.(7),

$$c_{i+1} > c_i$$

Fig.8 shows a simulation corresponding to the "duplication" experiment shown in Fig.3 in which a graft is introduced which contains sequences already present in the host. During intercalary regeneration these excess sequences are formed again, but with a reverted polarity. This leads to the correct neighbourhood of elements. This repair proceeds in the following steps. At the site of the discontinuity, the activity of the lower genes becomes dominant over that of the higher genes. In the example given, the area in which gene 2 is active increases until gene 3 is induced by the gene 2-activated cells, and so on, until the gap is repaired. In agreement with the experimental observations (Bohn, 1970; French, 1976) the model describes that cells from both sides participate

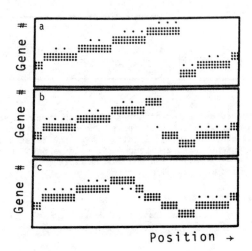

Fig.8: Simulation of intercalary regeneration (Eq.7). A gap is introduced in a sequence of feedback loops (genes) by juxtaposition of excessive elements, in analogy to the experiment shown in Fig.3. At the
gap, the distal elements (high gene number) have no support from their
normal proximal neighbours. They become reprogrammed into more proximal elements. If sufficient cells of a particular proximal state are
present, the induction of the next following state can take place as
shown in Fig.7. In this way, the gap becomes repaired. (After
Meinhardt and Gierer, 1980)

in the intercalation. No cell division is assumed in this calculation,
but it may be introduced into the model in a straightforward manner. Biological observation indicates that a gap stimulates cell divisions, pro-
viding more cells to form the missing structures.

13. GENERATION OF A SEQUENCE OF STRUCTURE BY DETERMINATION OF THE TER-
MINAL STRUCTURES AND A GENERAL DISCONTINUITY SIGNAL. In the preceeding
paragraph, a specific structure has been assumed to induce the chain of
adjacent structures. An alternative possibility would be that one or both
terminal structures are determined by a pattern forming process (Fig.1,
Eq.1) and that the remaining structures are induced by a signal which is
formed whenever celltypes which are normally not neighbours become juxta-
posed (Meinhardt and Gierer, 1980). Eq.8 describes an example for kinetic interactions with these properties. Due to the self-inhibitory term
d_i in eq.(8), the size of a particular element adapts to the total
size of the field.

$$\frac{\partial g_i}{\partial t} = \frac{c_i g_i'^2}{d_i \cdot r} - \alpha g_i \quad \text{with } g_i' = g_i + \delta g_{i-1}/p \qquad (8a)$$

$$\frac{\partial d_i}{\partial t} = \gamma(g_i - d_i) + D_d \frac{\partial^2 d_i}{\partial x^2} \qquad (8b)$$

$$\frac{\partial r}{\partial t} = \sum \frac{c_i g_i'^2}{d_i} - \beta r \qquad (8c)$$

$$\frac{\partial p}{\partial t} = \frac{rp}{q} - \mu p + P_o \qquad (8d)$$

$$\frac{\partial q}{\partial t} = rp - \nu q + D_q \frac{\partial^2 q}{\partial x^2} \qquad (8e)$$

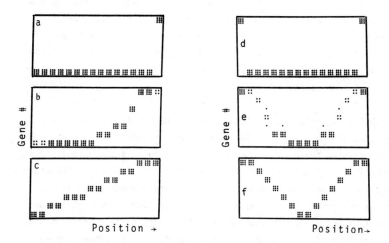

Fig.9: Gap repair and generation of a sequence by a general discontinuity signal (Eq.8). (a-c) Assumed is the determination of one terminal element (a). The gap in the sequence leads according Eq.8 to a low p-concentration (not shown) at the low site of the gap. This causes a sequential activation of higher feedback loops until the gap is repaired. New gaps are formed at neighbouring positions (b) which are repaired in the same way until each group of cells has its correct neighbours. Due to the self-inhibitory terms of Eq.8, the available space is shared evenly between the elements of the sequence. (d-f) terminal elements at both sides lead to two complete sequences; each element is present twice (f). Due to the self-inhibitory terms, each element occupies half as much space as compared with the normal situation (c).

Eq.8a,c is closely related to Eq.4, enabling the activation of a particular gene and the repression of the alternative genes. The steady
state concentration of the repressor r is proportional to c_i. Therefore
if $c_i < c_{i+1}$, the repressor concentration is a stable analog label –
a positional value – indicating which gene is active. Juxtaposition of
normally nonadjacent cells lead to a discontinuity in the r-concentra-
tion. This can be detected by the p-q system of Eq.8d,e which is essentially a linear version of the pattern formation equation (1a,b). It
leads to an increase of the p-concentration on one site and a decrease on
the other site of the discontinuity. The local drop of the p-concentration is assumed, according to Eq.8a, to initiate the repair of the gap.
Fig.9a demonstrates different stages in the generation of a sequence of
structures by gap repair, starting from one terminal structure. Further,
if the same terminal element is induced on both sites of the field, the
available space is evenly shared between the two sequences and two complete sequences are formed (Fig.9b). This proportion regulation is
caused by the long-range self-inhibitory terms d_i (Eq.8). Note that
this behavior is very different from those types of positional information schemes as discussed in section 3 were two terminal source regions
leads to a loss of structures inbetween.

14. CONCLUSION. Relatively simple interactions can lead to a reliable
generation of the manifold of structures in a developing organism. Sequences of structures can either be generated by more global positional
information signals or by chains of induction of neighbouring structures.
Mathematical equations accounting for such interactions allow a more
quantitative comparison of hypothesis and experimental observation and
assure that the assumed mechanisms are free of internal contradictions.
The properties of systems of non-linear feedback-loops often turn out to
be counterintuitive, requiring mathematical analysis rather than verbal
discourse for understanding. Further, we hope that a theoretical understanding will help to dissect the biochemical steps involved in pattern
formation.

BIBLIOGRAPHY

Bohn, H. (1965). Analyse der Regenerationsfähigkeit der Insektenextremität durch Amputations- und Transplantationsversuche an Larven der afrikanischen Schabe Leucophaea maderae Fabr. (Blattaria).II.Mitt. Achsendetermination. W.Roux' Arch. Entwickl.-Mech. Org. 156, 449-503.

Bohn, H. (1970). Interkalare Regeneration und segmentale Gradienten bei den Extremitäten von Leucophaea-Larven (Blattaria). I. Femur und Tibia. W. Roux' Archiv 165, 303-341.

French, V. (1976). Leg regeneration in the cockroach, Blattella germanica I. Regeneration from a congruent tibial graft/host junction. Wilh. Roux' Arch. 179, 57-76.

French, V. (1978). Intercalary regeneration around the circumference of the cockroach leg. J. Embryol. exp. Morph 47, 53-84.

Gierer, A. (1981). Generation of biological patterns and form: Some physical, mathematical and logical aspects. Prog. Biophys. Molec. Biol. 36, 1980.

Gierer, A., Meinhardt, H. (1972). A theory of biological pattern formation. Kybernetik 12, 30-39.

Gierer, A., Meinhardt, H. (1974). Biological pattern formation involving lateral inhibition. Lectures on Mathematics in the Life Science 7, 163-183. Providence, Rhode Island: The American Mathematical Society.

Kalthoff, K., Sander, K. (1968). Der Entwicklungsgang der Missbildung "Doppelabdomen" im partiell UV-bestrahlten Ei von Smittia parthenogenetica (Diptera, Chironomidae). Wilhelm Roux' Archiv 161, 129-146.

Meinhardt, H. (1977). A model for pattern formation in insect embryogenesis. J. Cell Sci. 23, 117-139.

Meinhardt, H. (1978a). Models for the ontogenetic development of higher organisms. Rev. Physiol. Biochem. Pharmacol. 80, 48-104.

Meinhardt, H. (1978b). Space-dependent cell determination under the control of a morphogen gradient. J. theor. Biol. 74, 307-321.

Meinhardt, H. (1980). Cooperation of compartments for the generation of positional information. Z. f. Naturforsch. 35c, 1086-1091.

Meinhardt, H., Gierer, A. (1974). Application of a theory of biological pattern formation based on lateral inhibition. J. Cell Sci. 15, 321-346.

Meinhardt, H., Gierer, A. (1980). Generation and regeneration of sequences of structures during morphogenesis. J. theor. Biol. 85, 429-450.

Nüsslein-Volhard, C. (1977). Genetic analysis of patternformation in the embryo of Drosophila melanogaster. Wilhelm Roux' Arch. 183, 249-268.

Rau, K. G., Kalthoff, K. (1980). Complete reversal of antero-posterior polarity in a centrifuged insect embryo. Nature 287, 635-637.

Sander, K. (1975). Pattern specification in the insect embryo. In: Cell Patterning. Ciba Foundation Symp. 29, pp. 241-263. Amsterdam: Associated Scientific Publishers.

Sander, K. (1976). Formatin of the basic body pattern in insect embryogenesis. Adv. Insect Physiol. 12, 125-238.

Schmidt, O., Zissler, D., Sander, K., Kalthoff, K. (1975). Switch in pattern formation after puncturing the anterior pole of Smittia eggs (Chironomidae, Diptera). Dev. Biol. 46, 216-221.

Spemann, H. (1938). Embryonic Development and Induction. New Haven: Yale University Press.

Wolpert, L. (1969). Positional information and the spatial pattern of cellular differentiation. J. theoret. Biol. 25, 1-47.

Wolpert, L. (1971). Positional information and pattern formation. Curr. Top. Dev. Biol. 6, 183-224.

MAX-PLANCK-INSTITUT
FÜR VIRUSFORSCHUNG
SPEMANNSTRASSE 35
D-74 TÜBINGEN
WEST GERMANY

Lectures on Mathematics in the Life Sciences
Volume 14, 1981

CONTROL OF OVULATION NUMBER IN A MODEL OF

OVARIAN FOLLICULAR MATURATION [1]

H. Michael Lacker   and   Charles S. Peskin

ABSTRACT. We assume that interactions between
developing follicles occur through circulating
hormones that control cell growth. This leads
to a system of ordinary differential equations
of the form

$$d\xi_i/dt = f(\xi_i, \xi) \quad \text{where} \quad \xi = \sum_i \xi_i$$

and where ξ_i measures the maturity of the i-th
follicle. We give a particular choice of f
for which the stable trajectories of this system correspond to a limited range of ovulation
numbers. Along these stable trajectories, the
population of follicles is divided into two
parts: a few follicles mature to ovulation
and the rest atrophy and disappear. For some
parameter values there are also stable trajectories that lead to pathological, anovulatory
states. The model takes on a probabilistic
aspect when we add the assumption that entry
into the class of interacting follicles occurs
by a stochastic process from a reserve pool of
immature, noninteracting follicles. We simulate
this situation numerically, and we compute the
distribution of ovulation times and ovulation
numbers.

Acknowledgments

    The authors would like to acknowledge the very signifi-
cant contributions made to this work by Jürgen Moser and
Jerome K. Percus. We would also like to thank Connie Engle
and Janice Hardware for typing the manuscript.

1980 Mathematics Subject Classification.
        92A05, 92A09, 34C35, 34F05.
1 Supported in part by the National Institutes of Health
  GM07552-03.

1. Introduction

   As the female, in most classes of animals, has
two ovaria, I imagined that by removing one it
might be possible to determine how far their
actions were reciprocally influenced by each other
... .   There are two views in which this subject
may be considered. The first, that the ovaria,
when properly employed, may be bodies determined
and unalterable respecting the number of young to
be produced... .   The second view of the subject
is, by supposing, that there is not originally
any fixed number which the ovarium must produce,
but that the number is increased or diminished
according to circumstances; that is it is rather
the constitution at large that determines the
number; and that if one ovarium is removed, the
other will be called upon by the constitution to
perform the operations of both, by which means
the animal should produce with one ovarium the
same number of young as would have been produced
if both had remained [1].
                              John Hunter, 1787

   In mammals the number of offspring produced in a litter
is usually characteristic of the species or breed. This is
a reflection of the relatively constant number of eggs that
are periodically shed from the ovaries at the time of ovulation.
   Each egg is released from a developmental unit called a
follicle which matures in the ovary over a period of weeks.
Follicles initiate growth continually from a large reserve
pool which is formed at birth. Only a few of those follicles which start to grow in each cycle actually mature
and release ova; the rest atrophy and die. The observed
variation in ovulation number for a given mammal is sufficiently small to rule out the hypothesis of independent
follicle growth. In fact, follicle interaction occurs
through circulating hormones. It is our hypothesis that this
interaction determines the number of follicles that
eventually mature to ovulation.
   The assumption of interaction through the circulation
is important. It will impose a particular symmetry on the
class of dynamical systems that can be used to describe

follicle growth. We shall use the word _global_ to describe
this kind of interaction.

2. Support for Global Interaction

The first evidence suggesting the possibility that global interaction regulates ovulation number appeared in 1787.
The famous Scottish surgeon John Hunter removed an ovary
from a sow in order to determine what effect this might have
on the size of its litters (see the quote that introduces
this paper). He found that although the reproductive life
span was significantly reduced by this manipulation, the size
of each litter did not change. Further examination showed
that the number of eggs released by the remaining ovary
doubles at each ovulation. If the operation is performed
at the right time in the cycle, compensation by the remaining ovary will occur at the next scheduled ovulatory period
[2,3]. This compensation has been observed in many mammals
and is called the law of follicular constancy [4].

If ovulation number is controlled only by local interactions between follicles, then the number of eggs released
at ovulation by one ovary should be independent of the
number released by the other. One would expect the removal
of one ovary to reduce the ovulation number by one-half.
However, if ovulation number is controlled by follicle interactions that occur through the circulation then any two given
follicles will interact in the same way whether they are
nearest neighbors or in different ovaries. Removing one
ovary should not change the number of eggs released at
ovulation provided that the size of the developing population of follicles is still large compared to the number
which eventually ovulate in a cycle.

Since global interaction is spatially independent, it
cannot be sensitive to the way a given set of follicles is
distributed between the two ovaries. This is unlikely to
be true if local, spatially dependent, interaction is
important. If follicles interact primarily through the
circulation then the distribution of eggs shed between the
two ovaries when conditioned on a given total ovulation

number should satisfy binomial statistics. If local interaction plays an important role then some deviation from the
binomial law might be expected. In fact, no significant
deviation from binomial statistics is observed in those
species which have been examined [5-10]. In mice, where this
distribution has been most extensively tested, p is very
close to 1/2.

It should be noted that binomial statistics and the law
of follicular constancy would also be satisfied if follicles
did not interact at all. However, as previously mentioned,
the assumption of independent follicle growth cannot explain
the small variation in ovulation number that mammals can
achieve. For example, the number of follicles activated per
cycle in a young woman is on the order of 10^3. Since the mean
ovulation number is one, the assumption of independent follicle growth leads to a prediction of 2 eggs being released in
18% of her cycles. One would think that this would lead to a
higher rate of fraternal twins than the observed rate of
about 1% [11].

3.   The Physiological Mechanism which Mediates
Global Interaction

Developing follicles might communicate directly through
their own secretions or they might interact indirectly by
controlling the release of growth mediating hormones from a
distant site. The latter possibility would appear to be more
compatible with a global mechanism and is, in fact, supported
by a large body of evidence. The distant site is the pituitary, whose secretions also help regulate other endocrine
glands.

Pituitary removal arrests follicle maturation in its
early stages. A chemical fraction from the gland, called
gonadotropin, can make follicles mature to ovulation in both
sexually immature animals and in animals whose pituitaries
have been removed. Gonadotropin consists of 2 protein
hormones called follicle-stimulating hormones (FSH) and
luteinizing hormone (LH). Specific receptors with extremely
high affinity for each of these hormones are found on follicle cells. The number of receptors a follicle contains

appears to depend on its maturity [12]. Both FSH and LH are
secreted in different amounts throughout the cycle. During
most of the cycle the ratio of FSH/LH is greater than 1,
however, in the few hours which precede ovulation there is
an abrupt increase in gonadotropin and the ratio reverses.

Indirect communication between follicles is established
by the fact that steroid secretions from developing follicles
regulate the release of gonadotropin. In several species,
the principal steroid hormone which regulates gonadotropin
release during follicle growth is estradiol. As a follicle
develops morphologically its estradiol secretory rate
increases [13,14]. The different cell types within a follicle cooperate in producing estradiol from cholesterol. In
addition to its ability to regulate gonadotropin release,
estradiol is a potent stimulator of cell division within a
follicle. Many of the effects of FSH and LH on follicle
development may be mediated through estradiol and its pre-
cursors [15]. These local effects may explain the influence
of a follicle's maturity on its own growth rate.

On the time-scale of follicle maturation (days), estradiol is rapidly removed from the circulation. The rate
of removal is proportional to the concentration with a time-
constant measured in minutes [16,17]. Thus equilibrium is
rapidly achieved. The resulting serum concentrations of
estradiol are at least an order of magnitude lower than those
concentrations measured in follicular fluid [18].

4. Follicle Maturation

Mammals are born with a large reserve pool of immature
follicles that decays exponentially with age [19]. Once a
follicle leaves this pool it will either ovulate or atrophy.
Each follicle in the immature pool consists of an egg cell
which is surrounded by a small number of granulosa cells
(order 10^1). A basement membrane separates the granulosa
cells from the remaining ovarian tissue.

No new follicles enter the reserve pool after birth and
follicles only leave by starting to grow. A growing follicle
can be recognized microscopically by an increase in the

number of its granulosa cells. The mechanism which triggers
granulosa cell division in a reserve follicle is not understood. Activation from the reserve pool continues in the
absence of gonadotropin although follicle maturation does
not proceed beyond its early stages. Exponential decay of
the reserve pool suggests that follicle activation may be
described by a poisson process.

   Each follicle is surrounded by a vascular net which
forms outside the basement membrane. In this area the
surrounding ovarian tissue differentiates to form an indis-
tinct shell called the theca. At the biochemical level
follicle maturation is a very complex process. Many inter-
connected reactions between granulosa and theca appear to
unfold in a coordinated way that is just beginning to be
understood.

   At any given time there is a distribution of follicle
maturities in the ovary. In humans follicles range in size
from 10^{-2} cm in diameter to 2.5 cm. The largest follicles
contain 10^7 granulosa cells. Before puberty all growing
follicles atrophy at different times and stages of growth.
Periodically, after puberty, a small and remarkably constant
number, depending on the species, complete maturation and
release their eggs nearly simultaneously on the time scale
of follicle growth. In mice, where the most careful measurements have been made, the estimated time from growth initiation to ovulation is 3 weeks [20].

5.   Formulation of the Model

   In this section, a model is proposed to describe the
interaction of developing follicles by means of circulating
hormones. For the sake of clarity, very simple and specific
physiological assumptions are made. In fact, however, the
model is not critically dependent on all of these assumptions.
A more general derivation has been given in (21).

   Consider a population of N developing follicles. Each
follicle will be characterized by the number of granulosa
cells $X_i(t)$, $i = 1,2,\ldots,N$. We will assume that follicle

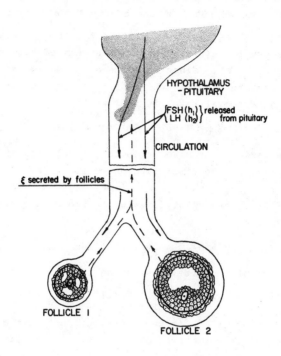

Figure 1. Schematic representation of the interaction
between 2 developing follicles. Follicle estradiol
secretory rate is used as a measure of follicle
maturity. The circulating concentration of estradiol, ξ , is assumed to control the release of the
pituitary gonadotropins FSH and LH. These pituitary hormones regulate the rate of follicle maturation. However, the response of a follicle to the
circulating concentrations of FSH(h_1) and LH(h_2)
at any particular time is also assumed to depend on
follicle maturity. (Reprinted from [21].)

extradiol secretory rate is proportional to X_i although any increasing monotonic relationship between these two quantities
could be used without altering the form of the model that is
eventually proposed. Let σ be the constant of proportionality.

Assume that estradiol is distributed in the serum volume
V at concentration $\xi(t)$ and that it is removed from the
serum at a rate $\gamma \xi$. Since the rate of change of serum
estradiol must be equal to the difference between its production rate and removal rate, it follows that

$$V \frac{d\xi}{dt} = \sum_{i=1}^{N} \sigma X_i(t) - \gamma \xi \,. \qquad (1)$$

If estradiol is removed from the circulation at rates
which are fast on the time scale of follicle growth [22-24]
then $\xi(t)$ is always near its equilibrium value. More pre-
cisely, if $X_i(t)$ are slowly varying on the time scale given
by $\tau = V/\gamma$, then

$$\xi(t) = \frac{\sigma}{\gamma} \sum_{i=1}^{N} X_i(t) \,. \qquad (2a)$$

It is convenient to write (2a) in the form

$$\xi(t) = \sum_{i=1}^{N} \xi_i(t) \qquad (2)$$

where $\xi_i(t) = \sigma X_i(t)/\gamma$ is the contribution that the ith
follicle makes to the estradiol concentration at time t.
Since $\xi_i(t)$ is proportional to $X_i(t)$ it is also a measure of
follicle maturity.

We assume that serum estradiol regulates the pituitary
production of LH and FSH. As above, we assume that the equi-
libration rates for the circulating concentrations of
these hormones are fast compared to the time scale of follicle maturation [25-27]. Then we can define the functions

$$\begin{aligned} h_1: \; &\xi \longmapsto \; h_1(\xi) \\ h_2: \; &\xi \longmapsto \; h_2(\xi) \end{aligned} \qquad (3)$$

where h_1 and h_2 are the circulating concentrations of FSH
and LH.

Finally, we assume that the specific growth rate of
granulosa cells in a follicle depends on the concentrations
of circulating gonadotropins and the maturity of the

follicle. That is

$$dX_i/dt = X_i\overline{\Phi}(X_i,h_1,h_2) \quad . \tag{3a}$$

X_i appears explicitly on the right-hand side to emphasize
the underlying exponential character of cell growth. The
function $\overline{\Phi}$ can be given a simple physiologic interpretation
as follows: If for any given follicle we define the net
growth rate as the net difference between the rate of cell
division and the rate of cell death, then $\overline{\Phi}$ is the net
growth rate per cell, or the specific growth rate.

Equation (3a) can be rewritten in terms of ξ_i as
follows

$$d\xi_i/dt = \xi_i\phi(\xi_i,\xi) \tag{4}$$

where we have introduced the function ϕ defined by

$$\phi(\xi_i,\xi) = \overline{\Phi}(\tfrac{\gamma}{\sigma}\,\xi_i,h_1(\xi),h_2(\xi)) \quad . \tag{5}$$

Our model consists of equations (2) and (4) which we
write together as a system for future reference:

$$\left.\begin{aligned} d\xi_i/dt &= \xi_i\phi(\xi_i,\xi) \quad i = 1,\ldots,N \\ \xi &= \sum_{j=1}^{N} \xi_j \end{aligned}\right\} \tag{6}$$

It is important to recognize that the effects of FSH
and LH on follicle growth are still present in equation (6).
They are represented implicitly through equations (3) and (5).
The system (6) really represents a class of models which
becomes a particular model when ϕ is specified, as we shall
do.

There are two important symmetries in (6). First, the
form of ϕ is the same for all i. This means that all follicles satisfy the same law of growth. The actual growth
rate may differ in different follicles at the same time
because the value of the growth rate depends on maturity ξ_i.
The second symmetry in (6) is that interaction between follicles occurs only through ξ which is a symmetrical
function of the ξ_i. This is an expression of the assumption
that follicle growth is regulated by global interactions
that are exerted through circulating hormones. All follicles

with identical maturity ξ_i are assumed to respond in the same
way when exposed to a given circulating hormonal environment.
At any time the environment is controlled by ξ.

6. The Growth Law Function ϕ

Since ϕ is arbitrary, it might seem that the class of
models represented by the system (6) is too broad to be
useful. In fact, as we have just shown, the symmetries in
(6) are very restrictive. Is it possible for a model to
exhibit the correct qualitative features of follicle maturation given the restrictions that all follicles have the same
program for development and interact only through the
circulation? Such a model should possess the following
properties:

(1) It should allow a few follicles to emerge from the
the developing population with ovulatory maturity while the
remainder atrophy and die at different times and stages of
growth;

(2) The number of ovulatory follicles should be relatively constant and emerge at regular intervals even though
follicles start growing at random times;

(3) It should be able to account for the fact that
mammalian species and breeds have different characteristic
ovulation numbers.

A priori, it is not clear whether a growth law ϕ exists
which will generate this qualitative behavior. We shall
answer this question by giving an example. Whether or not
this example actually corresponds to the developmental program of follicles in the ovary remains to be seen.

The specific example which we will analyze is:

$$d\xi_i/dt = \xi_i \, \phi(\xi_i,\xi) \ , \quad i = 1,\ldots,N$$

$$\xi = \sum_{j=1}^{N} \xi_j \tag{7}$$

$$\phi(\xi_i,\xi) = 1 - (\xi-M_1\xi_i)(\xi-M_2\xi_i)$$

where M_1 and M_2 are constants that remain unchanged for
each i.

At this point the reader who is familiar with the biochemical complexities involved in the hormonal regulation of
cell growth may protest that the model is clearly too simple
to be realistic. The answer to this objection is that the
growth law ϕ is not intended as a detailed description of all
the processes involved in the regulation of follicle maturation. Instead, ϕ is supposed to summarize the relevant
consequences of the biochemistry. In fact, we have not
arrived at ϕ by studying the biochemistry but by a process
of unnatural selection in which various growth laws were
tried out and their predictions compared with the behavior
of the ovary. A major task for the future is to relate the
growth law to the underlying biochemistry.

The particular model represented by the system (7) is
motivated by the stability of its equilibria and the properties of its symmetric solutions. These will now be
discussed.

Consider the behavior of (7) when M follicles have
exactly the same maturity and all others are dormant. Since
ξ is the sum of the contribution made by M identical follicles, it follows that

$$\xi_i(t) = \begin{cases} \xi(t)/M \, , & i = 1,\ldots,M \\ 0 & , \quad i = M+1,\ldots,N. \end{cases} \qquad (8)$$

When (8) is substituted into (7) the dynamics simplify to

$$d\xi/dt = \xi + \mu\xi^3 \qquad (9)$$

where $\mu = -(1- M_1/M)(1- M_2/M)$. Without loss of generality
we will assume $M_2 > M_1$.

When the M follicles are sufficiently small the linear
term in (9) dominates and the follicles grow independently
and exponentially. As the follicles grow the cubic term
begins to play a role. The role that it plays depends on the
number of interacting follicles M (Fig. 2).

If M is outside the interval (M_1,M_2), then $\mu < 0$ and the
follicles will approach an equilibrium maturity

$$\xi_M = 1/\sqrt{(M-M_1)(M-M_2)} \qquad (10)$$

independent of the initial maturity. This equilibrium is

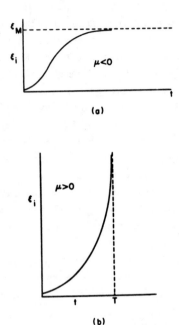

Figure 2. Qualitative behavior of the symmetric solutions of
equation (6). These correspond to the special
case where M developing follicles are imagined to
interact with <u>exactly</u> the same maturity and all
other follicles are assumed dormant. If the
number of follicles, M, is between M_1 and M_2 ($\mu > 0$),
then an ovulatory solution develops (2b). However,
if M lies outside the interval (M_1, M_2) then the M
follicles approach an equilibrium
$\xi_M = 1/\sqrt{(M-M_1)(M-M_2)}$ maturity (2a). Note that the
idealized ovulation time, T, (2b) depends on M and
initial maturity (see equation (11)).
(Reprinted from [21].)

stable within the framework of the symmetric solutions given by (8), but it usually becomes unstable when unsymmetric perturbations are considered (see below).

If on the other hand the number of identical follicles M is within the interval (M_1, M_2) then the interaction will be stimulatory ($\mu > 0$) and the M follicles will "ovulate" in a finite time given by

$$T = \frac{1}{2} \ln \left((1+\mu\xi_0^2)/\mu\xi_0^2 \right), \tag{11}$$

where ξ_0 is the initial value of ξ. Solutions which "blow up" in finite time are appropriate to represent ovulatory solutions for the following reason. In women and primates $\xi_i(t)$ for an ovulatory follicle has actually been measured during the later part of the follicular growth phase of the cycle. This is possible because the serum estradiol concentration at this time is almost entirely due to a single ovulatory follicle. The concentration does not approach an equilibrium but continues to increase in slope [28] even on a logarithmic scale. High, fast rising serum estradiol levels appear to be important in triggering the gonadotropin surge. On the time scale of follicle growth, this surge is essentially an instantaneous event that causes follicle rupture and egg release.

Thus, when M identical follicles interact, ovulation numbers are restricted to lie within the range (M_1, M_2). In reality follicles interact with different maturities because they start to grow at different times. However, at any instant, there is only one concentration of each circulating gonadotropin. This suggests that it might be instructive to consider the behavior of the function ϕ when ξ is fixed and ξ_i varies.

As a function of the maturity ξ_i the growth rate ϕ has a parabolic form with a maximum at a particular maturity

$$\xi_{i_{max}} = \frac{1}{2}(M_1 M_2 / (M_1 + M_2))\xi$$

that depends on the instantaneous value of ξ. When ξ_i differs too much in either direction from this optimal maturity the growth rate is negative. Thus the model promotes the growth of follicles whose individual maturities lie in a

certain range. Since the optimal maturity is proportional to
ξ a group of follicles will be selected for growth as ξ
increases in time. Of course, the estradiol concentration
develops in time as the net result of the simultaneous growth
and atrophy of all follicles in the interacting population.

If we assign to each of the N follicles a coordinate
axis in N-dimensional space, then any set of maturities can
be represented by a point P with coordinates $(\xi_1, \xi_2, \ldots, \xi_N)$.
The special solutions (8) of M identical follicles will lie
along the line of symmetry ℓ_M associated with each M-dimensional coordinate hyperplane (see Fig. 3a). Those lines of
symmetry in coordinate hyperplanes with dimension outside
the interval (M_1, M_2) will contain an equilibrium point P_M
which blocks ovulation. (Each P_M has M coordinates equal to
the equilibrium maturity $\xi_M = 1/\sqrt{(M-M_1)(M-M_2)}$, and the
remaining N-M coordinates equal to 0.) If M is within the
interval (M_1, M_2) then no equilibria will lie on ℓ_M. Any
starting point on one of these ℓ_M will escape along this line
to ∞ in a finite time given by equation (11). These correspond to ovulatory solutions.

7. Stability of Symmetric Equilibria

Additional insight about the behavior of the growth law
is obtained by examining its linearized behavior near each of
the equilibria P_M. Because of the symmetry in the problem it
is enough to consider the equilibrium whose first M coordi-
nates are ξ_M. Let $\tilde{P} = (\tilde{\xi}_1, \tilde{\xi}_2, \ldots, \tilde{\xi}_N)$ be a perturbation
from this equilibrium. Substituing $P_M + \tilde{P}$ into equation (7)
and assuming that $\tilde{P}$ is small produces the linear system.

$$\dot{\tilde{P}} = A\tilde{P}$$

$$A = \left(\begin{array}{ccc|ccc}
a_1+b_1 & & b_1 & & b_1 & \\
 & \ddots & & & & \\
b_1 & & a_1+b_1 & & & \\
\hline
 & & & a_2 & & 0 \\
 & & & & \ddots & \\
0 & & & 0 & & a_2
\end{array} \right) \begin{array}{c} \\ M \\ \\ \\ M \\ \\ \end{array} \qquad (12)$$

where $a_1 = \xi_M \frac{\partial\phi}{\partial\xi_i}\Big|_{\xi_M,M\xi_M}$, $a_2 = \phi(0,M\xi_M)$, and $b_1 = \xi_M \frac{\partial\phi}{\partial\xi}\Big|_{\xi_M,M\xi_M}$

are functions of M and the growth law parameters M_1 and M_2 (see Table 1). It should be noted that the symmetric equilibria P_M do not exhaust the set of stationary states of (7). Equilibria off the lines of symmetry also exist. They are all unstable and will not be analyzed in any great detail.

It is easily demonstrated that the eigenvalues of A are either the eigenvalues of the symmetric block in the upper left or the diagonal block in the lower right. Although there is a complete set of N independent eigenvectors for each P_M , there are only 3 distinct eigenvalues $\lambda_S = a_1 + Mb_1$, $\lambda_{in} = a_1$ and $\lambda_{out} = a_2$ (see Table 1).

The eigenvectors of A have a simple geometric interpretation which is most clearly illustrated by considering the following example (see Fig. 3a). Suppose the number of interacting follicles, N, is 3 and the growth law parameters of (7) are $M_1 = 1.9$ and $M_2 = 2.9$. The stability analysis suggests that, with probability 1, two follicles will ovulate and one will atrophy.

Since the interval (M_1,M_2) includes the integer 2, the lines of symmetry, ℓ_2 , in each 2-dimensional coordinate plane will be ovulatory. That is, if 2 follicles start with exactly the same maturity and the third is dormant, then both follicles will ovulate. However, since 1 and 3 are both outside (M_1,M_2), the coordinate axis, ℓ_1, and the line of symmetry in 3-space, ℓ_3 , will each contain a stationary point. These are labelled P_1 and P_3 in Fig. 3a. P_1 prevents one follicle from ovulating when the other 2 are dormant. P_3 prevents 3 follicles from ovulating when they all have the same maturity. The three eigenvectors of A at P_1 and P_3 will now be considered.

The eigenvector associated with $\lambda_S = a_1 + Mb_1$ lies along ℓ_M. From the behavior of the symmetric solutions (Fig. 2a) we expect $\lambda_S < 0$ for M outside (M_1,M_2). Table 1 shows that this is indeed true. There are N-M independent eigenvectors associated with $\lambda_{out} = a_2$. They correspond to perturbations from P_M which are out of the M-dimensional

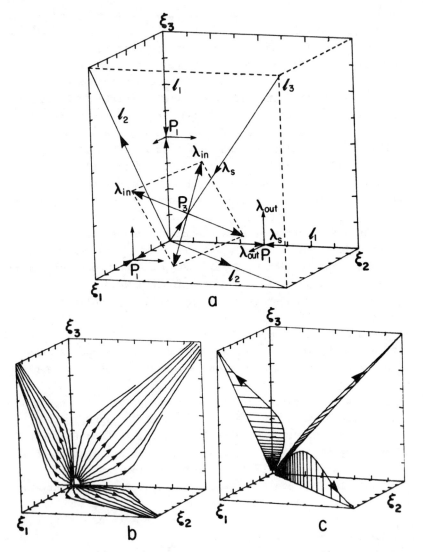

Figure 3. See text for explanation. The dashed lines in (a)
          are out of the plane of the paper. The origin is
          in the plane.   The curves in (b) are in the coordi-
          nate planes and those in (c) are in 3-space.

          (Reprinted from [21].)

coordinate hyperplane. For P_1 , there are $N - M = 2$ perturbations out of the 1-dimensional coordinate hyperplane (the coordinate axis). These are illustrated in Fig. 3a. $\lambda_{out} > 0$ for P_1 , since $1 = M < M_* = M_1 M_2/(M_1 + M_2) = 1.15$ (see Fig. 4 where λ_{out} is sketched as a function of M).

Fig. 3b illustrates that P_1 directs nearby solutions in the coordinate planes towards ℓ_2. In fact, all solutions in the plane will asymptotically approach ℓ_2 and "blow-up" in finite time. Thus two small nondormant follicles of unequal maturity will both grow and the smaller one will "catch-up" in finite time and ovulate.

The final eigenvalue $\lambda_{in} = a_1$ has an M-1 dimensional eigenspace that corresponds to perturbations from P_M that lie in the M-dimensional coordinate hyperplane but are orthogonal to ℓ_M. For P_1 no such space exists (the eigenvectors of P_1 are already complete). For P_3 , M-1 = 2. The perturbations from P_3 which lie in the 2-dimensional space orthogonal to ℓ_3 are shown in Fig. 3a. $\lambda_{in} > 0$, because $M = 3 > 2.9 = M_2$ (see sketch of λ_{in} as a function of M in Fig. 4. P_3 directs all nearby solutions towards the 2-dimensional coordinate planes which contain ovulatory solutions (Fig. 3c). In fact all phase curves except ℓ_1 and ℓ_3 asymptotically approach ℓ_2 and "blow-up" in finite time. Since any small perturbation from ℓ_1 and ℓ_3 will approach ℓ_2 in finite time, two of the three interacting follicles will ovulate and one will atrophy and die.

If the growth law parameters are kept the same as in the example above but the number of interacting follicle N is made larger than three, then the stability analysis suggests that all but two follicles will atrophy and die. In this case, all the stationary states P_M in higher dimensional coordinate hyperplanes than three are unstable saddle point equilibria which direct nearby solutions toward lower dimensional coordinate subspaces (see Table 1 and Figs. 4 and 5). This suggests that all phase curves will asymptotically approach the 2-dimensional coordinate hyperplanes which are filled with ovulatory trajectories.

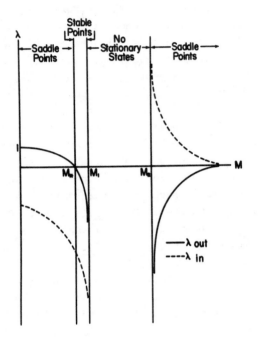

Figure 4. Sketch of the eigenvalues of A (see (12)) as a
function of M (see Table 1). λ_{out} is associated
with those eigenvectors which are orthogonal to
the M-dimensional coordinate hyperplane. λ_{in} is
associated with perturbations which are orthogonal
to ℓ_M but within the M-dimensional coordinate
hyperplane. The eigenvalue λ_s associated with
the eigenvector along the line of symmetry ℓ_M
is always stable and is not indicated in the
diagram. Only integer values of M have physical
meaning.
(Reprinted from [21].)

TABLE 1
(Reprinted from [21].)

A Summary of the Stability Analysis for Equilibria P_M.

Eigenvalues of the variational matrix A	Eigen-value Multi-plicity	Perturbation Eigenvectors $\bar{Z}=(\delta\xi_1,\ldots,\delta\xi_N)$	Geometric Interpretation in N-Dimensional Phase Space of $\bar{Z}=(\delta\xi_1,\ldots,\delta\xi_N)$
$\lambda_s = a_1 + Mb_1$ $= -2$	1	$\bar{z}_s$ satisfies $\delta\xi_i = \begin{cases} 1, & i=1,\ldots,M \\ 0, & i=M+1,\ldots,N \end{cases}$	A perturbation from the stationary point P_M along ℓ_M.
$\lambda_{in} = a_1 =$ $\dfrac{(M_1+M_2)M-2M_1M_2}{(M-M_1)(M-M_2)}$	$M-1$	$\bar{z}_1,\ldots,\bar{z}_{M-1}$ independent vectors which satisfy $\sum_{i=1}^{M}\delta\xi_i = 0,$ $\delta\xi_i=0 , \ i=M+1,\ldots,N$	Any perturbation from the station-ary point P_M perpendicular to ℓ_M but within the M-dimensional coordinate hyperplane.
$\lambda_{out} = a_2 =$ $1 - \dfrac{M^2}{(M-M_1)(M-M_2)}$	$N-M$	$\bar{z}_{M+1},\ldots,\bar{z}_{N-1}$ satisfy $\delta\xi_i = 0, \ i=1,\ldots,M$ $\sum_{i=M+1}^{N}\xi_i=0, \ i=1,\ldots M$ and $\bar{z}_N$ satisfies $\delta\xi_i=\begin{cases}(M-N)b_1, & i=1,\ldots M \\ (a_1-a_2)+Mb_1, & i=M+1,\ldots,N\end{cases}$	Perturbation out of the M-dimensional coordinate hyper-plane.

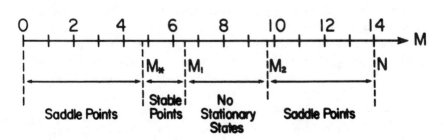

Figure 5.   Stability of equilibria P_M located on lines of symmetry ℓ_M. (Reprinted from [21].)

Of course, ovulatory solutions need not be restricted to
a single ovulation number. M_1 and M_2 can be chosen so that
several coordinate hyperplanes will contain ovulatory solutions (those M-dimensional coordinate hyperplanes in the
interval (M_1,M_2)). In addition, for special values of M_1 and
M_2 it is possible for some of the equilibria P_M to become
asymptotically stable. As shown in Fig. 4 all eigenvalues
are negative when M is between $M_* = M_1M_2/(M_1+M_2)$ and M_1.
Integers in this interval will correspond to stable P_M.

Each of these stable equilibria has a domain of attrac-
tion, and when the initial condition lies within such a
domain, ovulation will not occur. Instead M follicles will
become stuck at an equilibrium maturity given by equation
(10). Since these "stuck" follicles continue to secrete
hormones, they can be a source of additional pathology in
the uterus and breast where cell growth is regulated by
steroids secreted from the developing follicles.

8. Numerical Solutions

In this section we will test and further develop our
intuition about the behavior of the growth law by solving
the system (7) numerically. The numerical method used to
obtain these results will be described after we discuss the
change of variables on which it is based (Section 9).

Initial conditions are determined in the following way.
Each of N follicles is independently assigned a starting
maturity that is chosen at random from a uniform distribution
on the interval $(0,\xi_{max}^0)$. The observation that the reserve
pool decays exponentially with age suggests that follicle
activation could be modeled as a poisson process. For now,
however, we will consider each cycle to begin with N follicles activated at the same time but with different maturities.

Figure 6 shows the results of 4 cycles in which 10
follicles are activated in each cycle. Although each follicle
satisfies the same growth law ($M_1 = 3.85$, $M_2 = 15.15$ for
each follicle), some follicles continue to mature while
others atrophy and die. The results of many cycles, with

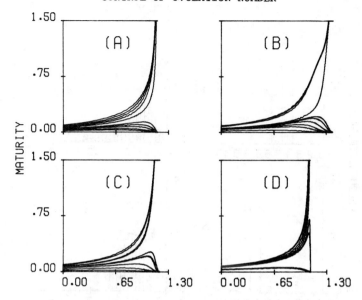

Figure 6. Follicle maturation curves in 4 cycles. Each curve
represents the development of a follicle whose
initial maturity is chosen at random from a uniform
distribution of maturities between 0 and 0.1.
Although every follicle obeys the same law of
growth, some follicles are selected for continued
development while others become atretic. The growth
growth law parameters M_1 and M_2 of equation (7) are
are the same for each follicle ($M_1 = 3.85$, $M_2 = 15.15$).
In cycles (A) and (D) 5 ovulatory follicles emerge.
In (B) and (C) the ovulation number is 4. In each
cycle 10 follicles interact. Note that it is
possible for an ovulatory follicle and an atretic
follicle to have almost the same maturation curve
for most of the length of the cycle (see cycle D).
On the other hand, a significantly smaller follicle
can occasionally "catch-up" and ovulate (see cycle
B). The ovulation time is slightly different in
each cycle.
(Reprinted from [21].)

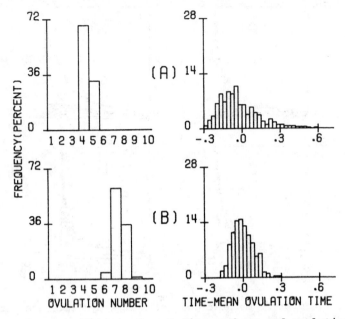

Figure 7. Distribution of ovulation numbers and ovulation
          times. In (A) the growth law parameters are the
          same as for Fig. 6 (M_1 = 3.85, M_2 = 15.15). The
          statistics are obtained for 500 cycles. In every
          cycle 30 follicles interact. Each follicle is given
          an initial maturity that is chosen independently
          from a uniform distribution in the interval
          (0, 0.075). In (B) the growth law parameters have
          been changed to M_1 = 5.5, M_2 = 61.7. The results
          summarize 300 cycles. As in (A), 30 follicles
          interact in each cycle but the initial maturities
          are uniformly distributed in the interval (0, 0.02).
          Statistics (mean ± SD): (A) ovulation number =
          4.32 ± 0.47, ovulation time = 1.39 ± 0.23;
          (B) ovulation number = 7.34 ± 0.58, ovulation time
          = 1.12 ± 0.08.

the same parameters as above, show that two ovulation numbers
are highly favored (Fig. 7). Even though ovulatory solutions
have been shown to exist for all integers in the interval
(M_1, M_2), the larger ovulation numbers are not observed at
all! This surprising result will be explained in Section 9.
Figure 6 could be interpreted in terms of a threshold initial
maturity which separates ovulatory from atretic follicles.
However, the results indicate that this threshold is
"automatically adjusted" in each cycle so that number of
follicles which ovulate is nearly independent of the initial
maturities. Figure 7 also shows the distribution of ovulation times which is unimodal and skewed in favor of shorter
intervals. This qualitative shape is observed in many
species including humans.

    The distribution of ovulation numbers is independent of
ξ_{max}^0. The shape of the ovulation time distribution is
unaffected by the choice of ξ_{max}^0 so long as follicles start
at maturities dominated by the independent exponential growth
phase.

    An interesting and perhaps important physiologic feature
of the model occurs when the size of the interacting population changes. Figure 8 shows a shift in the distribution of
ovulation numbers towards lower integers in the interval
(M_1, M_2) as the size of the interacting population increases.
There is also striking improvement in the control of ovulation time as N increases. These results suggest that the
large number of follicles which initiate growth but atrophy
and die in each cycle are playing an important role in the
regulation of ovulation number and time.

    The distribution of ovulation number seems to converge
with increasing N to some limiting distribution. The probability density of ovulation times appears to have a singular
limit with the standard deviation decreasing by a factor of
$1/\sqrt{N}$. Although each ovulation number has its own distribution of ovulation times (Fig. 9), nevertheless all of these
conditional distributions appear to converge to the same
singular limit as $N \to \infty$. This means that the limiting cycle.
time is independent of the ovulation number.

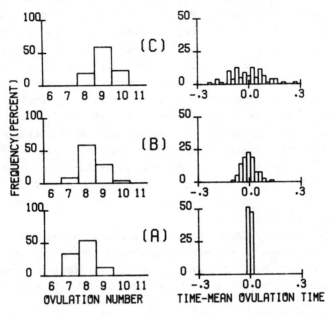

Figure 8. The effect of interacting follicle population size
on the distribution of ovulation times and numbers.
Larger numbers of interacting follicles improve
control of ovulation time and favor smaller ovulation numbers. In (A) 1000 follicles interact in
each cycle. In (B) 100 follicles interact per
cycle and in (C) 30 follicles interact. Each graph
represents the results of 80 cycles. Every follicle obeys equation (7) with M_1 = 6.1, M_2 = 5000.0.
Initial maturities are chosen at random to be a
number from a uniform distribution between 0 and
10^{-5}. Statistics (mean $\pm$ S.D.): (A) ovulation
number = 7.79 $\pm$ 0.65; ovulation time = 4.37$\pm$0.01;
(B) ovulation number = 8.82 $\pm$ 0.67, ovulation time
= 5.55 $\pm$ 0.04; (C) ovulation number = 9.04 $\pm$ 0.65,
ovulation time = 6.33 $\pm$ 0.10.
(Reprinted from [21].)

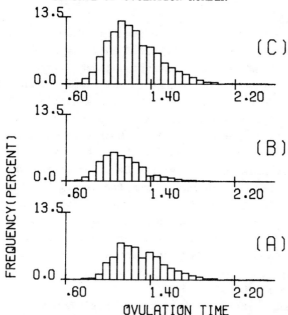

Figure 9. The distribution of ovulation times conditioned on
ovulation number. The initial maturity of each
follicle is chosen independently from a uniform
distribution in the interval (0, 0.05). The growth
law parameters (M_1 = 5.5, M_2 = 61.7) are the same
as for Figure 7 (B). In each cycle the number, N,
of interacting follicles is 10. (A) represents
the ovulation time frequencies for those cycles in
which 7 follicles ovulate (mean $\pm$ S.D. = 1.31$\pm$0.25).
(B) represents the distribution of ovulation times
for those cycles in which 8 follicles ovulate (mean
$\pm$ S.D. = 1.13 $\pm$ 0.20). The area under each graph
is equal to the probability of achieving that ovula-
ovulation number. The results are obtained from a
total of 1500 cycles. The distribution of ovulation times for all cycles is represented in (C)
(mean $\pm$ .S.D = 1.25 $\pm$ 0.25).
(Reprinted from [21].)

Figure 10 illustrates the results of 4 cycles when the growth law parameters are M_1 = 6.5 and M_2 = 15.5. As predicted from the stability analysis, these parameters admit the possibility of both ovulatory solutions and anovulatory states, since the interval (M_*, M_1) contains the integers 5 and 6. In Fig. 10B, 6 follicles become "stuck" at the predicated equilibrium maturity. Much more infrequently 5 follicles (Fig. 10D) become "stuck" (2 out of 1000 trials).

We now briefly consider the behavior of the growth law when follicles are activated from the reserve pool at random times. In Fig. 11 (preliminary results), follicles initiate growth at random times given by a poisson process. Each activated follicle is given the same initial maturity and obeys the same growth law (equation (7) M_1 = 3.15, M_2 = 15.15. Note that the number of interacting follicles, N, is now a function of time.

Although there is no source of periodicity, a relatively constant number of follicles periodically emerge as ovulatory. Just before ovulation the serum estradiol concentration is almost entirely due to the ovulatory follicles. At ovulation, these follicles are removed from the interacting population. This results in a precipitous drop of circulating estradiol to levels where stimulatory interactions can again occur and allow a new crop of follicles to mature. The ovulation number and time need not be the same as the previous cycle because of the assumed stochastic nature of growth initiation from the reserve pool.

An intriguing property of this form of the model is that successive cycles are completely uncorrelated. That is, any two random variables (e.g. ovulation times) associated with different cycles are independent. This follows from the fact that, in our model, atretic follicles are driven to ξ_i = 0 (zero rate of estradiol production) at the moment of ovulation, so they are completely removed at the same time as the ovulatory follicles. Thus the process has no memory from one cycle to the next. This raises the question whether successive cycles are correlated or not (the answer may be different in different species). Some preliminary evidence

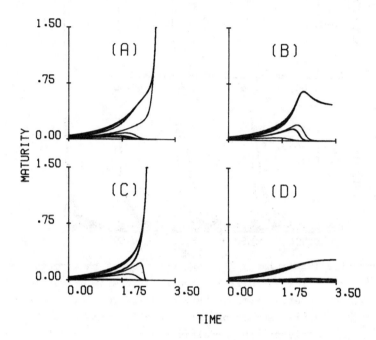

Figure 10. Follicle maturation curves for parameters which
admit both ovulatory solutions and anovulatory
states. Every follicle satisfies equation (7)
with the same parameter values M_1 = 6.5, M_2=15.5.
The initial maturity of each follicle is chosen
at random from a uniform distribution in the
interval (0, 0.05). In (A) and (C) 7 follicles
ovulate. In (B) an anovulatory state occurs in
which 6 follicles approach an equilibrium
maturity of 0.46. In (D) 5 follicles approach a
maturity of 0.25. Note that the approach to
equilibrium need not be monotonic (B).

(Reprinted from [21].)

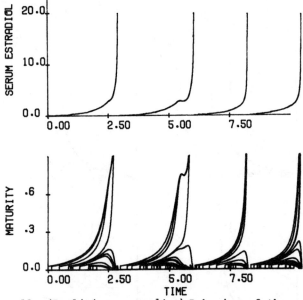

Figure 11. (Preliminary results.) Behavior of the growth law
(7) when follicles begin to develop at random
times determined by a poisson process. Each small
vertical mark on the time axis represents the
activation of a follicle. All activated follicles
start with the same maturity (0.03) and obey the
same law of growth.   (M_1 = 3.85, M_2 = 15.15).
Although there is no source of periodicity, a
relatively constant number of follicles mature at
periodic intervals.   The results of the 4
illustrated cycles are tabulated below:

Cycle	Ovulation Number	Ovulation Time	Follicles Activated per Cycle
1	4	2.91	18
2	4	3.12	27
3	5	2.20	17
4	5	2.21	20
TOTAL	18	10.44	82.

The mean activation rate is set at 8.

bearing on this question is discussed in [29].

In this idealized model ovulation number and time are determined entirely by interactions between activated follicles. These interactions are exerted indirectly through the effects of serum estradiol on the release of circulating pituitary hormones. In several species other factors besides secretions from developing follicles influence gonadotropin release. Some of these are in the external environment and periodically change with the season or day (light). Other factors include steroid production by sources other than growing follicles including, for example, the corpus luteum. As previously mentioned these factors cannot alone account for the small variance in ovulation number that mammals achieve. They can, of course, modulate the mechanism proposed for the regulation of ovulation number and time and therefore should be considered in more detailed schemes.

9. Stability of N-Space Trajectories

The stability analysis of symmetric equilibria suggested that the phase curves in N-space would approach the lines of symmetry, ℓ_M , in coordinate hyperplanes with dimensions in the interval (M_*, M_2). It was shown that these lines either contained ovulatory solutions (for M between M_1 and M_2) or stable anovulatory equilibria (for M between M_* and M_1). Although the numerical results agree with these theoretical predictions, in the sense that all of the observed values of M fall within the allowed interval, the larger ovulation numbers in this interval are not observed even though ovulatory solutions for these numbers exist. It might be thought that these ovulation numbers simply have a low probability of occurring. In fact, however, the larger allowed ovulation numbers have probability zero. As we now show, this is because ovulatory solutions in the interval $(2M_*, M_2)$ are unstable.

Since ovulatory phase curves blow up in finite time, a change of variables is chosen in which these curves approach finite equilibria whose stability can be analyzed by the methods employed in Section 7. This is accomplished

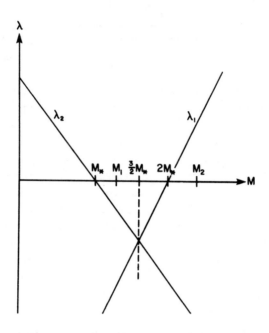

Figure 12. Stability analysis of the equilibria represented
by equation (16). Each equilibrium is characterized by M, the number of follicles with relative
maturity $\gamma_i = 1$. λ_1 and λ_2 are the distinct eigenvalues associated with each equilibrium. These
eigenvalues are linear in M (see equation (17))
and divide the range of ovulation numbers in the
interval (M_1, M_2) into a stable range $(M_1, 2M_*)$ and
an unstable range $(2M_*, M_2)$. This explains why
larger ovulation numbers in the interval (M_1, M_2)
are not observed even though solutions for these
ovulation numbers have been shown to exist (see
Section 6). The figure also shows a stable region
between $M_* = M_1 M_2 / M_1 + M_2$ and M_1. If integers
exist in this interval they correspond to the
presence of stable anovulatory states.

(Reprinted from [21].)

in the following way which was suggested to us by J. Moser.

Arrange the N follicles in order of their maturities with ξ_1 the most mature. (Since the $\xi_i(t)$ do not cross, the order of maturities is preserved in time.) We now rescale the time by defining

$$\tau(t) = \int_0^t \xi_1^2(t') \, dt' . \tag{13}$$

As t approaches the finite time of ovulation, $\tau \to \infty$. This follows from the fact that $d\xi_1/dt \sim \xi_1^3$ as $\xi_1 \to \infty$. Since the inverse of $\tau(t)$ exists we can use it to define the variables.

$$\gamma_i(\tau) = \xi_i(t(\tau))/\xi_1(t(\tau)) \quad \text{and} \quad \Gamma(\tau) = \xi(t(\tau))/\xi_1(t(\tau)). \tag{14}$$

We can now rewrite the system (7) in the following form.

$$\left. \begin{array}{c} d\gamma_i/d\tau = \gamma_i \, \psi(\gamma_i,\Gamma), \quad i = 1, N \\[2mm] \Gamma = \displaystyle\sum_{j=1}^{N} \gamma_j \end{array} \right\} \tag{15}$$

where

$$\psi(\gamma_i,\Gamma) = (1-\gamma_i)[M_1 M_2(\gamma_i+1) - \Gamma(M_1+M_2)].$$

Note that $\gamma_1(\tau) \equiv 1$, and $0 \le \gamma_i(\tau) \le 1$.

Since we expect $\xi_i(t)/\xi_1(t) \to 1$ for ovulatory follicles we look for stationary points of (15) of the form

$$\gamma_i = \begin{cases} 1, & i = 1,\ldots,M \\ 0, & i = M+1,\ldots,N . \end{cases} \tag{16}$$

In fact, this form is also applicable to anovulatory states, and the only way to tell the difference is to see whether ξ_1 is infinite or finite. The stability analysis of (15) near these stationary states is summarized in Fig. 12. There are only 2 distinct eigenvalues λ_1 and λ_2. They are linear in M.

$$\lambda_1 = (M_1+M_2)M - 2M_1 M_2$$
$$\lambda_2 = -(M_1+M_2)M + M_1 M_2 \tag{17}$$

Since ovulatory solutions correspond to integers between M_1 and M_2 , we see that this interval is broken into a stable and unstable region. The stable ovulation numbers lie at the lower end between M_1 and the harmonic mean, $2M_*$, of M_1

and M_2. The larger ovulation numbers are unstable. The
region between M_* and M_1 is also   stable. As noted earlier
in Section 7, if integers exist in this region they correspond to stable anovulatory states.

The above analysis is consistent with the numerical
results obtained in Section 8. For example, consider Fig. 7B.
Although symmetric solutions for ovulation numbers between 6
and 61 exist, only ovulation numbers 6 through 9 are observed.
Since $M_1 = 5.5$, $2M_* = 10.1$, these results match the
stability analysis perfectly except that the ovulation number
10 should have been seen. Presumably its low probability is
related to the fact that 10 is very close to the stability
boundary in this case.

It should be noted that the range of stable ovulation
numbers is not affected by the number of interacting follicles since the eigenvalues λ_1 and λ_2 are independent of N.
Changing N, however, alters the dimension of the phase space
and presumably the geometry of the capturing region associated with a given stable ovulation number. This could change
the frequencey with which a given stable ovulation number
will be observed.

10. Numerical Methods

The numerical solutions of Section 8 were not obtained
by direct integration of (7). Advantage was taken of the fact
that the solutions of (15)   do not "blow-up" in finite time
but rather approach finite stationary points asymptotically
as $\tau \to \infty$. The initial maturities $\xi_i(0)$ are converted to
relative maturities $\gamma_i(0) = \xi_i(0)/\xi_1(0)$. The system (15) is then
directly integrated by utilizing a standard explicit finite
difference method (a second order Runge-Kutta scheme was
used). The value of $\Gamma(\tau) = \sum_{i=1}^{N} \gamma_i(\tau)$ converges to an
integer which is the ovulation number (or number of folli-
cles stuck in an anovulatory state). Since the distribution
of relative maturities determines the ovulation number, it is
easy to understand why the distribution of ovulation numbers
is independent of ξ_{max}^0.

Once $Y_i(\tau)$ and $\Gamma(\tau)$ have been obtained, transformation to $\xi_i(t)$ is accomplished by solving the differential equation which is satisfied by $\bar{\xi}_1(\tau) = \xi_1(t(\tau))$,

$$\frac{1}{2}\frac{d}{d\tau}\bar{\xi}_1^2 = 1 - (\Gamma - M_1)(\Gamma - M_2)\,\bar{\xi}_1^{-2}. \tag{18}$$

The actual time t which corresponds to τ is obtained by integration

$$t(\tau) = \int_0^\tau \frac{1}{\bar{\xi}_1^2(\tau')}\,d\tau'. \tag{19}$$

Using the inverse $\tau(t)$ we finally obtain,

$$\xi_i(t) = \bar{\xi}_1(\tau(t))\,Y_i(\tau(t)). \tag{20}$$

The solutions obtained were checked against a scheme which solved the untransformed equations (7) directly. Both numerical methods converged to the same solution.

The advantages of using the change of variables are that the dependent variables Y_i are bounded while the ξ_i are not and that high time resolution is automatically achieved near the blow up time because $\tau \to \infty$ as t approaches the time of ovulation.

Conclusion

We have proposed a simple and specific model in which developing follicles regulate their growth through interactions that are exerted by circulating hormones. These interactions occur indirectly through follicle secretions that control the release of pituitary hormones. Although all follicles obey the same developmental program and start growing at random times from an immature reserve pool, a small and relatively constant number of follicles emerge at regular intervals with ovulatory maturity. The remainder atrophy and die at different times and stages of development. A change in the parameters of the growth law can alter the distribution of ovulation numbers and times. Thus, the observation that mammalian species and breeds have different characteristic litter sizes can be accounted for by the same basic developmental scheme.

An interesting and perhaps important physiologic
feature of the model occurs when the size of the interacting
population changes. As the number of interacting follicles
increases, there is striking improvement in the control of
ovulation time. There is also a shift in the distribution
of ovulation numbers towards smaller integers. Since the
number of interacting follicles in a cycle decreases with
age, these results partially explain the increased variance
in the time of ovulation as women approach menopause [30].
The behavior of the model is also consistent with the
observed increase in the occurrence of dizygotic twins as
women age [31].

The model therefore suggests two important functions
for the large number of follicles that are activated during
each cycle, even though an overwhelming fraction (about
99.9% in humans) of these are destined to atrophy and die.
The functions of these nonovulatory follicles are to hold
down the ovulation number and to reduce the variance in the
time required for the ovulatory follicles to mature.

This makes sense in evolutionary terms when we consider
that, in primitive species, the reproductive strategy is to
produce as many offspring as possible. Mammals have a
different reproductive strategy in which a large effort is
invested in a small number of offspring. Often in evolution,
older mechanisms are not discarded but are adapted to new
ends when the demands of the environment change. This is
often accomplished by superimposing a new layer of control
mechanisms upon an older scheme. In the case of the ovary,
the model suggests that the old strategy of producing as
many ova as possible has been adapted to the (opposite)
purpose of tight control on the reproductive process. Before
the present model was proposed, the significance of the
nonovulatory follicles was completely mysterious.

Another important feature of the model is that it
predicts, under special conditions, the existence of (pathological) anovulatory states. These have been shown to
correspond to stable equilibria in which a certain number of
follicles become "stuck" and produce nearly steady levels of

circulating estradiol. In some women it has been observed
that such states may exist for long periods of time [32].
It is important to understand such states because estradiol
and its metabolites are potent stimulators of cell growth
in the uterus and breast. Persistent exposure to steady,
relatively high levels of estradiol may have serious consequences including a greater risk for the development of
carcinoma of the breast and endometrium [33].

The duration of these states and their frequency of
occurrence in a given individual varies over a wide spectrum
in the female population [34]. Spontaneous escape does occur
and may be the result of random perturbations. Such perturbations could occur naturally, for example, by the continual
and random entry of follicles into the interacting population
from the reserve pool. The factors in the model which influence the duration and frequency of occurrence of these states
are presently being investigated. It should be noted that
these stable anovulatory equilibria exist only for special
values of the parameters. Therefore the model is consistent
with the observation that some species and some individuals
within a species do not exhibt these states.

Clearly, at the biochemical level, the mechanisms that
regulate the growth of follicles are far too complicated to
be described by any equation as simple as our growth law ϕ.
The question, however, is not whether this growth law contains a detailed description of all of the processes involved
but whether it is an adequate summary of the relevant consequences of these complicated biochemical events. One
important test of this is whether the growth law generates
behavior consistent with observations on the control of ovulation number. Another important test is whether the growth
law is actually obeyed by individual follicles. The first
test has already been passed, as demonstrated in this paper.
Experiments designed to test the second point are proposed
in [21].

References

[1]   Hunter, J. (1787). An experiment to determine the effect
      of extirpating one ovarium upon the number of young
      produced. Philos. Trans. R. Soc. Lond. Ser. B.$\underline{77}$, 233.

[2]   Greenwald, G. S. (1961). Quantitative study of follicu-
      lar development in the ovary of the intact or uni-
      laterally ovariectomized hamster. J. Reprod. Fertil. $\underline{2}$,
      351.

[3]   Peppler, R. D. and Greenwald, G. S. (1970). Effects of
      unilateral ovariectomy on ovulation and cycle length in
      4- and 5-day cycling rats. Am. J. Anat. $\underline{127}$, 1.

[4]   Lipschütz, A. (1928). New developments in ovarian
      dynamics and the law of follicular constancy. Br. J.
      Exp. Biol. $\underline{5}$, 283.

[5]   Falconer, D. S., Edwards, R. G., Fowler, R. E., and
      Roberts, R. C. (1961). Analysis of differences in the
      numbers of eggs shed by the two ovaries of mice during
      natural estrous and after superovulation, J. Reprod.
      Fertil. $\underline{2}$, 418.

[6]   McLaren, A. (1963). The distribution of eggs and
      embryos between sides in the mouse. J. Endocrinol. $\underline{27}$,
      157.

[7]   Brambell, F. W. (1935). Reproduction in the common
      shrew, I. The estrous cycle of the female. Philos. Trans.
      R. Soc. Lond. Ser. B $\underline{225}$, 1.

[8]   Brambell, F. W., and Hall, K. (1937). Reproduction of
      the lesser shrew. Proc. Zool. Soc. London, 106, 957.

[9]   Brambell, F. W. and Rowlands, I. W. (1936). Reproduc-
      tion of the bank vole, Vol. I. The estrous cycle of the
      female. Philos. Trans. R. Soc. Lond. Ser. B $\underline{226}$, 71.

[10] Danforth, C. H., and deAberle, S. B. (1928). The
      functional interrelation of the ovaries as indicated by
      the distribution of fetuses in mouse uteri. Am. J. Anat.
      $\underline{41}$, 65.

[11] Parkes, A. S. (1976). "Patterns of Sexuality and Repro-
      duction," Oxford Univ. Press, London, p. 79.

[12] Richards, J. S. and Midgley, A. R., Jr. (1976). Protein
      hormone action: a key to understanding ovarian follicu-
      lar and luteal cell development. Biol. Reprod. $\underline{14}$, 82.

[13] Baird, D. T. (1977). Synthesis and secretion of
      steroid hormones by the ovary in vivo. In "The Ovary",
      Vol. 3 (2nd Ed.) (S. Zuckerman and B. T. Weir, Eds.),
      Academic Press, New York, pp. 359-412.

[14] McNatty, K. P. (1978). Cyclic changes in antral fluid
      hormone concentrations in humans. In "Clinics in Endo-
      crinology and Metabolism", Vol. 7, No. 3 (G. T. Ross
      and M. B. Lipsett, Eds.), pp. 577-599.

[15] Ross, G. T. and Lipsett, M. B. (1978). Hormonal corre-
     lates of normal and abnormal follicle growth after
     puberty in humans and other primates. In "Clinics in
     Endocrinology and Metabolism", Vol. 7, No. 3 (G. T.
     Ross and M. B. Lipsett, Eds.), pp. 561-575.

[16] Tapper, C. M., and Brown-Grant, K. (1975). The secre-
     tion and metabolic clearance rates of estradiol in the
     rat. J. Endocr. 64, 215.

[17] Baird, D. T., Horton, R., Longcope, C., and Tait, J. F.
     (1969). Steroid dynamics under steady state conditions.
     Recent Prog. Horm. Res. 25, 611.

[18] McNatty, K. P. (1978). Op. Cit.

[19] Jones, E. C., and Krohn, P. L. (1961). The effect of
     hypophysectomy on age changes in the ovaries of mice.
     J. Endocrinol. 21, 497.

[20] Pedersen, T. (1970). Follicle kinetics in the ovary of
     the cyclic mouse. Acta Endo. 64, 304.

[21] Lacker, H. M. (1981). The regulation of ovulation number
     in mammals: an interaction law which controls follicle
     maturation. Biophysical Jour. 35 (In press).

[22] Baird, D. T., Horton, R., Longcope, C., and Tait, J. F.
     (1969). Op. Cit.

[23] Pedersen, T. (1970). Op. Cit.

[24] Tapper, C. M., and Brown-Grant, K. (1975). Op. Cit.

[25] Cargille, C. M., Ross, G. T. and Yoshimi, T. (1969).
     Daily variation in plasma follicle stimulating hormone,
     luteinizing hormone, and progesterone in the normal
     menstrual cycle. J. Clin. Endocrinol. Metab. 29, 12.

[26] Speroff, L., VandeWiele, R. L. (1971). Regulation of
     the human menstrual cycle. Am. J. Obstet. Gynec. 109,
     234.

[27] Tsai, C. C. and Yen, S. S. C. (1971). Acute effects of
     intravenous infusion of 17β-estradiol on gonadotropin
     release in pre- and post-menopausal women. J. Clin.
     Endocrinol. Metab. 32, 766.

[28] Baird, D. T. and Guevara, A. (1969). Concentration of
     unconjugated estrone and estradiol in peripheral plasma
     in nonpregnant women throughout the menstrual cycle,
     castrate and post-monopausal women and in men. J. Clin.
     Endocrinol. Metab. 29, 149.

[29] Winfree, A. (1980). The Geometry of Biological Time.
     Springer, New York.

[30] Korenman, S. G., Sherman, B. M. and Korenman, J. C.
     (1978). Reproductive hormone function: the perimenopausal
     period and beyond. In "Clinics in Endocrinology
     and Metabolism," Vol. 7, No. 3 (G. T. Ross and M. B.
     Lipsett, Eds.), W.B. Saunders Co., London, pp. 625-643.

[31] McArthur, N. (1954).   Statistics of twin births in Italy,
     Ann. Eugen. 17, 249.

[32] Speroff, L., Glass, R. H., and Kase, N. G. (1973).
     Clinical Gynecologic Endocrinology and Infertility
     (1st Ed.), Williams & Wilkins Co., Baltimore,
     pp. 61, 62.

[33] Speroff, L., Glass, R. H., and Kase, N. G. (1978).
     Clinical Gynecologic Endocrinology and Infertility
     (2nd Ed.), Williams & Wilkins Co., Baltimore,
     pp. 123-133.

[34] Yahia, C. and Taymor, M. L. (1970). Variants of the
     polycystic ovary syndrome. In "Meigs and Sturgis
     Progress in Gynecology," Vol. V (S. H. Sturgis and
     M. L. Taymor, Eds.), Grune & Stratton, New York,
     pp. 163-171.

COURANT INSTITUTE OF MATHEMATICAL SCIENCES
AND DEPARTMENT OF PATHOLOGY
NEW YORK UNIVERSITY   (H.M.L.)

COURANT INSTITUTE OF MATHEMATICAL SCIENCES
NEW YORK UNIVERSITY   (C.S.P.)

N.Y.U. Courant Institute of
Mathematical Sciences
251 Mercer Street
New York, New York 10012

Lectures on Mathematics in the Life Sciences
Volume 14, 1981

MODELING OF CELL AND TISSUE MOVEMENTS IN

THE DEVELOPING EMBRYO

S. Childress and J. K. Percus*

ABSTRACT.   Several models which have recently been used to
study morphogenetic movements are summarized. These have
the common feature of utilizing a unit representing a cell
or group of similar cells, as the basic building block in
both analysis and numerical simulation. We focus here on
the use of distributed interfacial tension as a means of
creating a representative stress field over a cell aggregate, which can be realized by varying the adhesive properties of cells. A free energy is then defined over a space
of allowable configurations. The path through configuration space representing a morphogenetic movement is then
realized by following the steepest descent of free energy,
under a constraint which fixes the rate of dissipation of
energy. The model is applied to cell monolayers, but the
approach may provide effective and economical simulations
in other geometries.

1. INTRODUCTION. Many of the events in the early development of

multicellular organisms involve the movements of cells or cell

aggregates. These morphogenetic movements, which are presumably

the consequence of instructions provided in the genetic material

of the cells, are part of the program of changes by which the em-

bryo organizes tissue material and gradually acquires structure

and form. Viewed objectively as a reorganization of material,

such movements can be modeled at a number of different levels,

depending on the detail of description of cell biology and

1980 Mathematics Subject Classification 73P05.
*
Supported by the National Science Foundation under
Grant MCS-79-02766 at New York University.

intercellular interaction. In this paper we examine some examples
of this modeling which are currently under study. Their common
feature is their reliance on an elemental unit, which may represent an individual cell, but the various approaches differ considerably in properties of these units and of their interactions.

Among the phenomena which fall into the category of morphogenetic movements we list the following: (1) The motion of individual
cells, and the forming and breaking of cell aggregates. (2) Rearrangement of cells within an aggregate. (3) Distortion of tissue,
e.g. invagination and evagination, and the thickening, bending,
and folding of cell sheets, which might be regarded as global responses to changes in the shape of individual cells. (4) Distortions produced by mitosis and the growth of tissue volume. In
approaching this vast repertory of intercellular activity, one can
envisage a broad spectrum of models, ranging from a detailed
analysis of cell microstructure to a continuum treatment of the
tissue material, and it is not clear that one should seek general
mechanisms applicable to more than one developmental event.

Some idea of the possible scope of models is provided by recent
studies of tissue distortion relying on the changes of shape of
constituent cells. In a now classic work, Jacobson and Gordon [1]
treated the elongation and thickening of the amphibian neural
plate in simulation, using a code which included a pattern of
cellular instructions determining cell height as a function of
time. These instructions were adjusted until the desired global
movements were achieved. This approach emphasizes the possibility
of exact simulation of some features of morphogenetic movements by
adjusting the system at the cellular level, and also indicates how
an "inverse method" for empirically determining developmental
strategies arises naturally in the present problem.

A more detailed model of cellular processes associated with
morphogenesis of the epithelium has recently been put forward by

Odell, Oster, Burnside, and Alberch [2]. In the model of Odell
et al. a cell monolayer is represented by a network of visco-
elastic elements, together with a certain number of contractile
elements. The contractile elements can be activated either by
stretching, or else biochemically. Once activated, epithelial
bending then becomes a cooperative activity in which a local program of cell responses is stimulated either mechanically or chemi-
cally, the global response depending upon the distribution of
contractile elements and the various parameters defining the
structure. This model opens up a number of exciting prospects
and the applications to gastrulation and neuralation discussed in
[2] yield a remarkable similarity to the observed events.

Another line of investigation, which we focus on in the present
paper, utilizes what is probably the simplest of cell interactions,
expressed in terms of some measure of bonding energy between contiguous cells. As we indicate in the next section, this admit-
tedly crude model of the cell leads naturally to a formulation of
"morphodynamics" (to use the term adopted by Jacobson and Gordon
[1]) involving a distributed field of interfacial tensions, where
"interface" is used generically to mean a boundary separating to
distinct cell types [3]. In some respects the tension elements
which arise here are similar to the elements utilized by Odell *et
al.* [2], and serve the same purpose, although there is no analog
of their activation by stretching.

2. DIFFERENTIAL ADHESION. On the basis of earlier work by Holtfreter [4] and Moscona [5], Steinberg has formulated a theory of
the equilibrium structures formed by heterogeneous aggregates of
cells [6]. In this model, the preferred structures are determined by the "adhesive energies" of binding between the constituent cell types. If N cell types are present (with the environment regarded filled by one cell type), the determining parameters are the densities E_{ij} , i,j 1,...,N measuring the bonding

energy per unit area of interface between cell type i and cell
type j. A total energy of adhesion of an aggregate of cell then
takes the form 3

$$E = \sum_{i>j} \sum_{j=1}^{N} I_{ij}T_{ij} \,, \qquad T_{ij} = \frac{1}{2}(E_{ii}+E_{jj}) - E_{ij} \qquad (1)$$

where I_{ij} is the area of the interface between cell types i and
j. A corresponding free energy of adhesion is then $V = K - E$
where K is an arbitrary constant. Steinberg postulates that a
locally stable equilibrium aggregate is one which locally maximizes E (or minimizes V) under perturbations of configuration.
From (1) it is seen that the problem of finding all locally stable
equilibria is equivalent to a discretized version of the classical
problem of arranging contiguous substances supporting interfacial
tensions T_{ij} so as to minimize the total free energy.

It is of interest to study how the parameters E_{ij}, defined
for any pair of cells, can be defined in terms of primitive parameters pertaining to a single cell. Steinberg [6] considered the
expression

$$E_{ij} = kn_in_j \qquad (2)$$

where k is a constant and n_i can be regarded as a density of bonding sites associated with the membrane of the ith cell. Another
choice is

$$E_{ij} = (k_i+k_j)\min(n_i,n_j) \qquad (3)$$

where k_i is an energy contributed by a site on the ith cell, and
n_i is again the corresponding site density. Roughly, (2) represents sparse sites fixed to the membrane, which bind when a site
pair lies within some target circle of fixed radius, while (3)
would correspond to floating sites, the bond density being determined by the cell with sparser sites. Kinetic bonding models
yielding (2) and (3) in special cases have been considered by
Bell [7].

If the number density of cells is large, so that the aggregate is roughly equivalent to a continuous material with smooth interfaces between cell types, the classical theory of hydrodynamic equilibria can be used to determine conditions insuring that cells of one type (A) will engulf cells of a second type (B), as shown in Figure 1. The necessary and sufficient conditions are

$$T_{AB} > 0, \quad E_{AB} > E_{AA} . \tag{4}$$

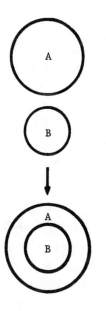

Figure 1. Engulfment of cell type B by cell type A.

If (4) are satisfied we write $A \supset B$. It is easily seen that for an arbitrary assignment of bonding energies E_{ij} the engulfment operation $\supset$ is not in general transitive, but it can be shown [6],[3] that transitivity holds for energies defined according to both (2) and (3). Note that without transitivity, there is no unique preferred "onion" configuration, obtained by sorting out within a spherical aggregate. If transitivity holds, the preferred layering is in the order of the engulfment sequence. Note also that the order, but not the configuration, is uniquely defined; for example, the engulfment shown in Figure 1 is complete once the aggregate of B-cells lies within the A-cells, both interfaces being spherical, but the equilibrium is indifferent to the location of the inner sphere.

For aggregates of finite cells the minimization of V can be

defined only in relation to an allowable set of cell configurations (e.g. a two-dimensional model might consist of a checker-
board of squares representing two or more cell types). But once
a set of permitted cell shapes and cell-cell contacts is defined,
it is possible to explore the path in the set of configurations,
which connects the initial and equilibrium states, by defining a
means of testing nearby configurations in a way that leads to a
decrease of V. A number of cell-sorting simulations have been
carried out based upon these general principles [8], and it has
been found that the method of testing neighboring configurations
can significantly bias the path. Unless relatively nonlocal permutations are permitted, it is possible for the sorting to ter-
minate at local equilibria which are far from the equilibrium of
absolute minimum V. Goel and Rogers [9] have overcome this difficulty by introducing a procedure for testing of configurations
which includes nonlocal permutations, which seems to simulate the
nonlocal pressure field in the analogous fluid flow problem.

Further elaboration of this class of models would seem war-
ranted, particularly in the search for dynamical models of movement which can be used in place of the essentially kinematic
approach to cell sorting. In the following sections we pursue
one such course, which attempts an extension of the equilibrium
theory, using as a general guide a well known procedure of non-
equilibrium thermodynamics. Starting from the variational principle of the equilibrium theory, we generate the path in configuration space by a steepest descent argument, using as constraint
some measure of the rate of dissipation in movement. We first
illustrate the model for the case of an isolated cell of simple
geometry.

3. THE SINGLE CELL. The interfacial surface tension isolated in
our first equation is presumably a consequence of the collective
interaction of a large number of cells. Strictly applied,

therefore, the concept only makes sense for large aggregates. On
the other hand, the internal stresses which change the shape of
individual cells can also be modeled most simply by surface
stresses of this kind. Greenspan [10] has applied this idea to a
mechanical model of cell division, and to the shape and movement
of fluid on a substrate [11]. In the present section we invoke
this approach in a simple geometry where the cell is two-dimensional and rectangular. Here and in the subsequent discussion
"cell" can actually refer to an aggregate of identical "sub-units"
which are not resolved in the model, with typical simulations
involving at most a few hundred of these cells.

Consider such a cell, of area A, resting on a graded substrate
(Figure 2). Let τ be the (constant) surface tension in the
exposed cell surface, and $\sigma(x)$
the time-independent tension at
point x of the cell-substrate
interface. Then a free energy of
the cell configuration (measuring
the work required to create surface with the assigned tensions),
is

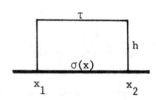

Figure 2. A single
rectangular cell on a
substrate.

$$V(x_1,x_2;\tau,\sigma) = \tau(x_2-x_1)$$
$$+ \int_{x_1}^{x_2} \sigma(x)\,dx + 2\tau A(x_2-x_1)^{-1} \quad (4)$$

An equilibrium configuration is obtained at a local minimum of V
with respect to the determining points $x_{1,2}$:

$$\frac{\partial V}{\partial x_1} = -\tau - \sigma(x_1) + 2\tau A(x_1-x_2)^{-2} = 0 \; ,$$

$$\quad (5)$$

$$\frac{\partial V}{\partial x_2} = \tau + \sigma(x_2) - 2\tau A(x_1-x_2)^{-2} = 0 \; .$$

At equilibrium, therefore, $\sigma(x_1) = \sigma(x_2) = \sigma$ and

$$(x_2-x_1)^2 = \frac{2\tau A}{\tau+\sigma}, \quad \text{or} \quad \frac{A}{x_2-x_1} = h_e = \left[\frac{(\tau+\sigma)A}{2\tau}\right]^{\frac{1}{2}}, \tag{6}$$

where h_e is the equilibrium cell height.

If σ is independent of x, then, up to a translation, the cell configuration is determined by the equilibrium cell height. The imposed rectangular shape thus predicts a vertically elongated cell when $\tau \ll \sigma$, and a cell whose base is twice its height when $\tau \gg \sigma$. In the theory of fluid equilibria with surface tension (zero gravity), the former case corresponds to an approximately spheroidal drop obtained when a fluid drop "rounds up" on a substrate, this being the manner in which contact with the substrate is minimized. That is, the extreme elongation of a rectangular cell is a result of the rectangular geometry. In the other case, where cell surface tension dominates, a liquid drop will assume a near hemispherical shape in order to balance horizontal tensions at points of contact with the substrate, so the rectangular geometry is in close correspondence.

Since, in the present example, a given configuration is completely determined by the points $x_{1,2}$, we refer to a configuration space $C(x_1,x_2)$. Morphogenetic movement will amount to motion through configuration space, and we wish to formulate an algorithm which uniquely determines such a motion from any initial condition. Because of the large effective viscosity of cell aggregates [12], and the low speed of observed morphognetic movements, one is led to consider a balance between the rate of decrease of free energy, and the rate of dissipation of energy into heat by unspecified dissipative processes. In the absence of any other reservoirs of internal energy (an example of possible importance would arise if the cell had some elasticity), this balance will be exact, leading to a monotonic decrease of free energy during the course of the movement. Starting from an initial configuration, we hypothesize that a given (infinitesimal) dissipation within the system is associated with that movement which maximizes the decrease of free energy. In other words, we assume that the path in configuration space corresponds to a steepest descent path on the surface $V(x_1, x_2; \tau, \sigma)$ defined over the configuration space. Note that this prescription defines movement in terms of an as yet unspecified rate of dissipation, determined in general by both coordinates and velocities in configuration space. Also, the overall rate of dissipation as a function of time will determine the instantaneous _speed_ of motion along the path. Using as an analogy the projection of a movie, a definition of rate of dissipation as a function of configuration will uniquely determine the sequence of frames on the film, but knowledge of the global dissipation as a function of time is needed to fix the projection speed at any instant. This kinematic description can be converted into a dynamical one, however, by adding the constraint that, along the path determined by steepest descent, the rate of decrease of free energy is in fact _equal_ to the rate of dissipation. This will yield dynamic equations on configuration space, and for the choices of dissipation function used below (suggested by viscous flow at low Reynolds number), involving quadratic dependence on the velocities, the procedure leads to a kind of viscous hydrodynamics driven by fields of surface tension.

To formulate the model in the simplest case consider the change in aspect ratio of a cell on an ungraded substrate σ = constant. Placing the center of the cell at the origin, the motion is symmetric in x. We adopt the dissipation Φ which would occur if the rectangle were filled with a viscous fluid of kinematic viscosity ν. With $u_i = \dot{x}_i$, we have

$$\Phi(u_1, u_2; x_1, x_2) = 4\nu A (u_2 - u_1)^2 (x_2 - x_1)^{-2} \quad . \tag{7}$$

To find the steepest descent path using (7), we consider the

68S. CHILDRESS AND J. K. PERCUS

variation

$$\delta[\frac{dV}{dt} + \lambda(t)\Phi] = \delta[u_i \frac{\partial V}{\partial x_i} + \lambda(t)\Phi] = 0 \qquad (8)$$

with respect to the velocities u_i, for fixed x_i and fixed
"time" t, $\lambda(t)$ being a multiplier. Thus

$$u_2 = -u_1 = (16\nu\lambda A)^{-1}[2\tau A - 4(\tau+\sigma)X_2^2] . \qquad (9)$$

Adding now the constraint

$$\frac{dV}{dt} + \Phi = 0 \qquad (10)$$

to determine $\lambda(t)$, we see that $\lambda = \frac{1}{2}$, a result that will apply
whenever Φ is a quadratic form in the velocities. The time
dependence of movement is now explicitly given by

$$x_2 = \left(\frac{\tau A}{2(\tau+\sigma)}\right)^{\frac{1}{2}} \tanh\left(\frac{\tau t}{4\nu} + c\right) , \qquad (11)$$

where c is a constant determined by the initial configuration.

On a graded substrate we can expect additional dissipation to
enter when a cell of fixed shape moves to a location of smaller σ.
We should then modify (7) through the addition of frictional dissipation, one possible choice being

$$\Phi = \frac{f}{4} (u_1+u_2)^2(x_2-x_1) + 4\nu A(u_1-u_2)^2(x_2-x_1)^{-2} , \qquad (12)$$

where f is a friction coefficient. Using (12) we generate a class
of motions which are analogous to the movements of a droplet calculated by Greenspan using the appropriate form of the viscous
flow equations [11].

We emphasize that the sequence of steps introduced above gene-
ralizes immediately to an arbitrary configuration space, free
energy, and dissipation function defined over a suitable set of
coordinates x_i and associated velocities u_i. A movement is

then generated by a choice of parameters in V and an initial configuration. The form of the relations (1)-(3) suggests that the
tension parameters be viewed as determined by two objects in contact. Thus, the parameter τ above is a contact tension between
cell and environment, σ a property of contact between cell and
an x-dependent substrate. Here and in the following sections we
neglect entirely the possible time-dependence of tension parameters, although this is an essential element in the initiation and
termination of movement. Some further discussion of this point is
given in Section 7.

4. MONOLAYERS. We consider now N adjacent cells of the preceding kind, lying on a graded substrate and carrying tensions e_i
on vertical interfaces, as indicated in Figure 3.* The free
energy is now defined by

$$V = e_0 h_1 + \sum_{i=1}^{N} \left[\int_{x_{i-1}}^{x_i} \sigma_i(x)\,dx + \tau_i(x_i - x_{i-1}) + e_i \min(h_i, h_{i+1}) \right.$$

$$\left. + T_i |h_{i+1} - h_i| \right] ,$$

$$T_i = \begin{cases} \tau_i, & h_i \geq h_{i+1} \\ \tau_{i+1}, & h_i < h_{i+1} \end{cases} ,$$

where e_0 and e_N must be chosen to properly reflect conditions

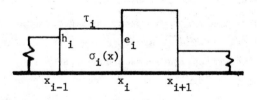

Figure 3. Monolayer of rectangular
cells.

---
*Note that we must have $\tau_i + \tau_{i+1} > e_i$ if cell-cell bonds are to
be maintained.

holding at the edges of the monolayer, and $h_{N+1} = 0$. If, for
example, the initial configuration is extended to all x as a periodic structure, $e_0 = e_N$ is arbitrary; if the right edge is free,
then $e_N = \tau_N$.

To compute the partial derivatives of V with respect to the x_i
it is helpful to introduce the variables $\xi_i = x_i - x_{i-1}$ and use

$$\frac{\partial V}{\partial x_i} = \frac{\partial V}{\partial \xi_i} - \frac{\partial V}{\partial \xi_{i+1}} . \tag{14}$$

Note that $\dfrac{\partial V}{\partial \xi_i}$ can be found by incrementing the ξ_j for $j > i$

the same amount. For example, with the configuration shown in
Figure 4 there results

$$\frac{\partial V}{\partial \xi_i} = \sigma_i(x_i) + \tau_i - \frac{h_i^2}{A_i} (e_i + \tau_i - \tau_{i+1}) . \tag{15}$$

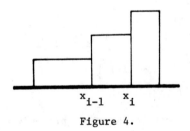

Figure 4.

Similar formulas apply for all other configurations. Note that
the right and left derivatives with respect to ξ_i may be different, although the exceptional cases do not seem to present a
problem in numerical simulation. The essential point is that the
geometry of the cells makes V only piecewise continuously differ-
entiable with respect to the x_i; discontinuous changes occur
when adjacent cells have the same heights. A related problem
occurs when there is a local extremum of cell height. We note
that, if $h_{i-1}, h_{i+1} < h_i$,

$$\frac{\partial V}{\partial \xi_i} = \sigma_i(x_i) + \tau_i - \frac{h_i^2}{A_i}\tau_i , \qquad (16)$$

which should be compared with (15). The different form will lead in simulations to fluctuating cell shape (see Section 8).

As dissipation function we may simply add contributions of the form (12):

$$\Phi = \sum_i \left[\frac{f}{4}(u_i+u_{i-1})^2(x_i-x_{i-1}) + 4\nu A_i(u_i-u_{i-1})^2(x_i-x_{i-1})^2\right] . \qquad (17)$$

To apply the model to the bending of cell sheets and the problems of morphogenesis studied by Odell *et al.* [2], we enlarge the configuration space to encompass the cells shown in Figure 5.

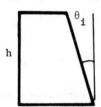

Figure 5. Cell geometry
with bending.

If $\theta_i \ll 1$ we can approximate V by

$$V \simeq \sum_i \left[\tau_i^{(1)}(\xi_i - h_i\theta_i/2) + \tau_i^{(2)}(\xi_i + h_i\theta_i/2) + e_i h_i\right] \qquad (18)$$

where $\tau_i^{(1,2)}$ are tension associated with the two segments of free membrane. The appropriate dissipation function, with $h_i\dot{\theta}_i$ and $u_i - u_{i-1} = \dot{\xi}_i$ occurring similarly, is seen to have the form

$$\Phi = \sum_i A_i\left[4\nu_1(u_i-u_{i-1})^2(x_i-x_{i-1})^{-2} + 4\nu_2 h_i^2\dot{\theta}_i^2(x_i-x_{i-1})^{-2}\right] . \qquad (19)$$

5. THE CONTINUUM LIMIT. An unusual property of the present models
is the reversal of conventional roles of discrete and continuum
representations. The genesis of structure at the cellular level
leads directly to numerical algorithms with the cell as "finite
element." Nevertheless the continuum limit can be useful for
clarifying the physical content of the model and for comparison
with other related physical problems.

To illustrate the idea we obtain here the continuum limit of
the monolayer model of Section 4. To accomplish this it is useful
to employ the Lagrangian variable

$$a_i = \sum_{j=1}^{i} A_j \qquad (20)$$

so that we may write $\Delta a_i = A_i$. It is then clear from (17) that
the continuum dissipation has the form

$$\Phi = f \int u^2 dx + 4\nu \int \left(\frac{\partial u}{\partial x}\right)^2 da , \qquad (21)$$

where a is now the continuum Lagrangian variable. Also, taking a
gradually sloping layer as typical, (14) and (15) yield

$$\frac{\partial V}{\partial x_i} = P_{i+1} - P_i \quad , \quad P_i = \frac{h_i^2}{A_i} (e_i + \tau_i - \tau_{i+1}) - \sigma_i(x) - \tau_i . \qquad (22)$$

We are thus led to introduce a representation of e_i as a difference. It is interesting that, if the tensions in (1) are computed
using (3) for nearby cell types, there results $T \sim k|\Delta n|$. (For
additional discussion concerning continuous cell types see [3].)
We thus suppose that a chemical potential μ exists such that
under the continuum limit $e = kd\mu$, so that from (22) the continuous variable replacing P_i is seen to be

$$P = h^2\left[k \frac{d\mu}{da} - \frac{d\tau}{da}\right] - \sigma - \tau . \qquad (23)$$

We thus consider the Euler–Lagrange equation obtained from the u-variation of

$$F = \int u \frac{\partial P}{\partial x} \, dx + \frac{1}{2} \, \Phi \; , \tag{24}$$

and there results

$$\frac{\partial P}{\partial x} - 4\nu \frac{\partial}{\partial x} h \frac{\partial u}{\partial x} + fu = 0 \; . \tag{25}$$

In the last expression we have used the Lagrangian form of the conservation of volume,

$$\frac{\partial x}{\partial a} = h^{-1} \; , \tag{26}$$

where $x(a,t)$ is the location at time t of the cell having label a. The Eulerian alternative to (26) has the well known form

$$\frac{\partial h}{\partial t} + \frac{\partial hu}{\partial x} = 0 \; , \tag{27}$$

where the time derivative is taken for fixed x.

Equations (25) and (27) must be supplemented by expressions which determine the parameters of V. If, for example, τ is a cell parameter (constant for fixed a), we would supply the function $\tau(a)$. The Eulerian expression of this property is clearly

$$\frac{\partial \tau}{\partial t} + u \frac{\partial \tau}{\partial x} = 0 \; . \tag{28}$$

For the case at hand it is reasonable to take τ, k , and μ as cell parameters; the property would presumably also apply to σ on an ungraded substrate.

It is not surprising that (25) has the form of the equation of motion of a shallow viscous layer driven by a pressure gradient, since this is largely a result of the form of the dissipation function, but the structure of the "pressure" function deserves scrutiny. If H_o is a typical height of the monolayer, and L_o

is its horizontal extent, then an appropriate dimensionless Lagrangian variable is $\alpha = a/H_o L_o$, and the dimensionless cell height is $h* = h/H_o$. Now if the height-to-width ratio of the cells is $0(1)$ or smaller, a large number of cells will force $\epsilon = H_o/L_o$ to be small, so that the effective pressure,

$$P = \epsilon h*^2 \left[k \frac{d\mu}{d\alpha} - \frac{d\tau}{d\alpha} \right] - \sigma - \tau \qquad (29)$$

contains a small parameter. This function includes contributions from both horizontal and vertical tensions. If both are to be comparable, then $kd\mu/d\alpha$ should be $0(\epsilon^{-1})$ and the term $\epsilon d\tau/d\alpha$ may be neglected, yielding

$$P \underset{\sim}{\sim} h*^2 k* \frac{d\mu}{d\alpha} - \sigma - \tau , \quad k* = k/\epsilon . \qquad (30)$$

We can also write this last equation in the form

$$P = k* \frac{d\mu}{d\alpha} (h*^2 - h*_e^2) , \qquad (31)$$

where $h*_e$ is the equilibrium height of the cell with label α. This unusual "equation of state" involves a quadratic dependence upon "density" $h*$, and a variable reference state $h*_e$. This form is similar to one suggested for use in the modeling of cell monolayers by Greenspan and Murray (private communication).

6. NUMERICAL STUDIES. A number of simulations of monolayer movement were carried out in order to test the above algorithm. We compared an "exact" algorithm, involving the computation of right and left derivatives of V with respect to the ξ_i as outlined in Section 4, with an "approximate" one using the effective pressures P_i , where

$$P_i = \frac{e_i}{A_i} h_i^2 - \sigma_i(x_i) - \tau_i . \qquad (32)$$

We also varied the relative contributions of frictional and

internal dissipation. The test problem represented the "squaring
up" of a layer on an ungraded substrate, the equilibrium configuration being one of twice the initial height, and the movement
being symmetric with respect to the origin (see Figure 6). For a
given iteration, velocities u_i were computed, and the time step
was adjusted so that the maximum change of cell height was a given
fraction of current cell height. The most striking feature of the
simulations was the tendency of cells to oscillate in shape about
some gradually changing mean value. This phenomenon, which is due
to the polygonal cell geometry and the normalization of time step
just described, is discussed in Section 8 below.

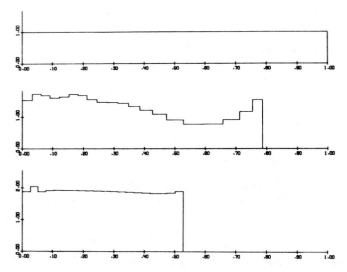

Figure 6. Monolayer movement using
the approximate algorithm with $\nu > 0$.

It was found that if no internal dissipation was present and
the approximate algorithm was used, the squaring up was extremely
slow and oscillatory behavior tended to dominate the cells' activity. This can be explained by the local nature of the frictional
dissipation; mechanical linkage between cells is in this case

essentially absent and the oscillation might be compared with capillary waves on a perfect fluid layer. Use of the exact algorithm did not significantly improve matters, and we conclude that some internal dissipation is essential for smooth movement. For all runs, however, some oscillations persisted, particularly near extrema of cell height, but the approximate algorithm compared satisfactorily with the exact one provided ν was sufficiently large.

To test the bending mode, we studied a problem treated by Odell *et al.* [2]: the symmetric distortion of an initially circular ring of cells due to differences in tension along the inner and outer surfaces. In these computations we used the approximate model corresponding to (18) and set $e = 0$. This problem involves constraints insuring that the band of cells remains intact, which can be conveniently introduced as side conditions on the variational problem in $\dot{\xi}_i$ and $\dot{\theta}_i$. Mechanically, these conditions introduce a distributed bending moment which provided an additional mechanical linkage between distant cells. Some representative computations are given in Figure 7.

7. REMARKS AND EXTENSIONS. The most important applications of these models may well be to problems in which cells do not necessarily maintain neighbors and for which the aggregate has a complicated two or three-dimensional form. D. Sulsky is currently applying the adhesion model to arrangements of cells in a two-dimensional monolayer. In this geometry, cells are assigned a fixed volume and are assumed to have the shape of right cylinders with polygonal boundary. The configuration space is determined by coordinates (x_i, y_i) of "nucleii," and cell boundaries are determined as a so-called Voronoi diagram using a procedure developed by C. Peskin [13]: each point of the plane is assigned to the nearest nucleus, as indicated in Figure 8. As cells move about, this construction allows bonds to be broken and formed, and

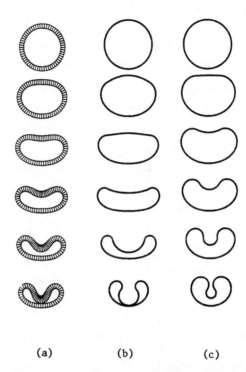

(a)   (b)   (c)

Figure 7.   Bending of a ring of cells due to tension on outer
boundary. (a) 60 cells with Gaussian distribution of tension.
(b) 196 cells with constant tension along upper 60. (c) Same
as (b) with tension confined to upper 30 cells. Computation
time approximately 10 sec CPU on the CDC 6600.

heights will change in response to changes in cross-sectional area.
The idea is to investigate how graded tensions, distributed over
the layer, determine the deformations of selected subgroups of
cells. Both frictional and internal dissipation is included. As
any given nucleus is virtually displaced, the cell boundary is
altered, changing locally the free energy, as determined by nearest neighbor parameters. In this way the cell can "explore"
nearby configurations of its boundary, and "nucleii" are used
primarily as a device to reduce the dimension of the configuration

space. Some preliminary applications to the neural plate problem
are now in progress, and it is hoped that it will be possible to
make a useful comparison between a simulation based on graded tension and the model developed by Jacobson and Gordon [1].

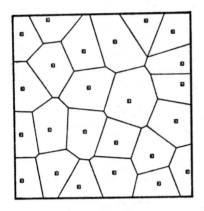

Figure 8. Cell membranes determined by "nucleii" according
to the principle "a point be-
longs to the nearest nucleus.

The activation and control of
morphogenetic movement involves
the union of mechanical processes
responsible for the changes in
aggregate geometry, with an underlying pattern of intercellular
communication. At the level of
the present modeling the latter
might take place through a diffu-
sible chemical, with the possibility suggested by Odell *et al.*
of a coupling between mechanical
stress and the transport of activator or inhibitor substances.

Greenspan [10] has emphasized the self-excitation of a stress
field which can occur when tensions are mediated by cell parameters, and there are a number of mechanisms by which movement can
be coupled to a diffusive instability of a subsidiary chemical
pattern [3]. Although these elaborations are at present highly
speculative, it is perhaps useful to distinguish between movements
which are activated by simply "turning on" a program, and those
which emerge through a bifurcation of structure, in the sense that
the system moves into a parameter domain allowing a preferred mode
of instability. In either case the resulting forms are likely to
be highly nonlinear functions of initial data (this is certainly
true of the models considered in this paper), and a systematic
exploration of this dependence could provide a kind of "calculus
of form" in specific models.

8. FLUCTUATIONS. In many examples of morphogenetic movement the motions at the cellular level are very erratic, and it is there-fore of interest that in the present monolayer simulations the naive numerical algorithms have just this property. Fluctuations are indeed desirable in any algorithm which presupposes a varia-tional definition of path, since in simulation this variability allows neighboring configurations to be continually "explored." To understand how these fluctuations arise it is helpful to ima-gine a sequence of one, two,..., N-cell problems as follows. Consider first a single cell in an ambient "bath" having fixed height h_o. The bath will have the parameters of the monolayer as a whole, which we take to constants e, τ, and σ, with $2\tau > e$.

Suppose now that, if the cell height exceeds h_o, its next value will be $(1-r)$ times the current value, while if the height is less than h_o the next value is $(1+r)$ times the current value, r being a positive number < 1. Then, the successive iterates will tend to oscillate erratically, at least for sufficiently large r. A similar situation occurs in the simulations near an extrema of cell height, since the derivatives of V at the extremal cell are very large compared to the values obtained from the gradual changes of cell parameters over a region of monotone increase or decrease of height. (In this sense the system is mathematically stiff.)

A similar analysis of a two-cell-plus-bath oscillation yields the phase portrait shown in Figure 9, in which we have let r have the value .65. It should be emphasized that the simulated fluctuations can be made as small as desired by increasing ν and decreasing r, and in that sense are not a serious problem. As already indicated, we regard explicit computations of the steepest descent as one means of introducing a random or at least fluctuat-ing component (albeit of a local nature) into the cell paths.

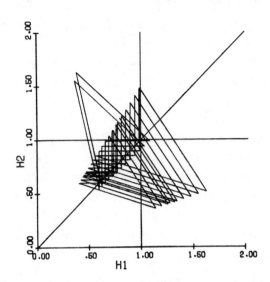

Figure 9. Cells of heights $h_{1,2}$ oscillating
in bath with mean height $h_o = 1$, $r = .65$.

9. STOCHASTIC DYNAMICS. In addition to the fluctuation which can
arise through the computational process, there will in general be
a certain "noise" in the system dynamics. Large local fluctuations are normal in biological processes. They arise from fluctuations in the large-scale parameters that determine the processes,
as well as from a profusion of fluctuating forces on a small scale
that are implied by the existence of macroscopic dissipation [14].
It is not easy to make a clean separation between these two
classes, or for that matter, between them and the effectively random impulses that are part and parcel of finite difference
numerical simulations of the type that we have considered. They
all have in common a high sensitivity to changes in initial data,
and the effect in each case is to replace exact dynamics by one
in which means, correlations, etc., are the more meaningful
characterizations.

Suppose that the dissipation function is given explicitly by

$$\Phi = \sum D_{ij}(h)\dot{h}_i\dot{h}_j \tag{33}$$

where h denotes the full set of h_i (e.g. cell heights together
with any h_0 needed to complete the specification of a configuration). According to (8) and (10), the dynamics in the absence of
fluctuations will be

$$\sum_j D_{ij}(h)\dot{h}_j = F_i(h) \tag{34}$$

where $F_i(h) = -\partial V/\partial h_i$ is the generalized force on h_i. A homo-
geneous set of uncorrelated (a severe assumption) random impulses
can then be applied in the form

$$\sum_j D_{ij}(h)\dot{h}_j = F_i(h) + f_i(h,t)$$

or
$$\dot{h}_i = \sum_j D_{ij}^{-1}(h)[F_j(h) + f_j(h,t)] \tag{35}$$

where the expectations

$$<f_j(h,t)> = f_j(h)$$

$$<(f_i(h,t) - f_i(h))(f_j(h',t') - f_j(h'))> = g_{ij}(h,h')\delta(t-t') \tag{36}$$

are specified. By standard techniques [15], the configuration
space probability distribution $\rho(h,t)$ then satisfies the
equation

$$\frac{\partial}{\partial t}\rho(h,t) + \sum \frac{\partial}{\partial h_i}[D_{ij}^{-1}(h)(F_j(h) + f_j(h)\rho(h,t)]$$

$$= \frac{1}{2}\sum \frac{\partial}{\partial h_i}\frac{\partial}{\partial h_j}[D_{ik}^{-1}(h)D_{j\ell}^{-1}(h)g_{k\ell}(h)\rho(h,t)] . \tag{37}$$

If the fluctuations are regarded as due to weak coupling and
consequent energy transfer to a thermal reservoir, classical

equilibrium statistical mechanics [16] tells us to expect an
equilibrium distribution

$$\rho(h) = e^{-\beta V(h)}/Q , \qquad (38)$$

where Q is the normalization and β an effective reciprocal
temperature. This requires a special choice of f_j and g_{ij},
converting (37) to

$$\frac{\partial \rho}{\partial t} - \sum \frac{\partial}{\partial h_i} [D_{ij}^{-1} \frac{\partial V}{\partial h_i} \rho] = \frac{1}{\beta} \sum \frac{\partial}{\partial h_i} [D_{ij}^{-1} \frac{\partial \rho}{\partial h_j}] . \qquad (39)$$

The kind of information we would like to obtain from (39) consists of the dynamics of the means and correlations of the h_i.
One common approximation proceeds by the ansatz

$$\rho(h,t) = e^{-\beta V(h,t)}/Q(t)$$

$$V(h,t) = V(h) + \sum \Delta_i(t)h_i \qquad (40)$$

through which any set $\langle h_i(t) \rangle$ can be enforced by suitable choice
of the $\Delta_i(t)$. The correlations $\langle h_i(t)h_j(t) \rangle - \langle h_i(t) \rangle \langle h_j(t) \rangle$
are then readily found from (40). For the $\langle h_i(t) \rangle$ themselves,
a mean field approach is then a possibility: it follows from (39)
that

$$\frac{d}{dt} \langle h_i \rangle = - \langle \sum_j D_{ij}^{-1} \frac{\partial V}{\partial h_j} \rangle + \frac{1}{\beta} \langle \sum_j \frac{\partial D_{ij}^{-1}}{\partial h_j} \rangle , \qquad (41)$$

and so if fluctuations are neglected,

$$\dot{h}_i = \sum_j D_{ij}^{-1} \partial V/\partial h_j + (1/\beta)\sum_j \partial D_{ij}^{-1}/\partial h_j . \qquad (42)$$

To the extent that the detailed nature of the stochastic forces on
the system is unimportant, this should, upon suitable choice of
the constant β , also correspond to the mean dynamics of a
numerical simulation.

10. QUASI-EQUILIBRIUM FLUCTUATIONS. Let us turn explicitly to the
monolayer of (13). A fair amount of information can be obtained
on the assumption that the system is locally uniform and in temporal equilibrium under the fictitious forces, e.g. the Δ_i in
(40), that produce the required local value of <h>. To avoid
end point problems, we imagine the system as periodically bounded.
Thus the relevant energy is

$$V = \sum_i \left[\frac{A(\sigma+\tau)}{h_i} + (e+\Delta)h_i \right] + \left(\tau - \frac{e}{2}\right) \sum |h_i - h_{i+1}| \qquad (43)$$

where $h_o \equiv h_N$. The computations involved in finding the normalization Q of (38) are then typical. We see that

$$Q = \int \dots \int e^{-\beta V(h)} \, dh^N = \text{Tr } T^N$$

where

$$T(h,h') = e^{-\lambda|h-h'|} \, e^{-\beta w(h')} \qquad (44)$$

$$\lambda = \beta\left(\tau - \frac{e}{2}\right) , \quad w(h) = \frac{A(\sigma+\tau)}{h} + (\ell+\Delta)h ,$$

and then in the same way

$$\langle h_i \rangle = \text{Tr}(T^N h)/\text{Tr}(T^N)$$

$$\langle h_i^2 \rangle = \text{Tr}(T^N h^2)/\text{Tr}(T^N) \qquad (45)$$

$$\langle h_i h_{i+1} \rangle = \text{Tr}(T^{N-1} hTh)/\text{Tr}(T^N) ,$$

where h denotes the diagonal operation of multiplying by h.

Now T is a non-negative operator of bounded norm on $[0,\infty]$,
and is similar to the Hermitian operator

$$\bar{T}(h,h') = e^{-\frac{\beta}{2} w(h)} T(h,h') e^{\frac{\beta}{2} w(h')} .$$

Therefore, T has a complete set of real discrete eigenvalues, its maximum eigenvalue t_o is non-degenerate, and the corresponding eigenfunction ψ_o non-negative. An immediate consequence is that for large N, it is only the maximal state (t_o, ψ_o) that contributes. Thus, as $N \to \infty$ (45) becomes

$$\langle h_i \rangle = (\bar{\psi}_o, h\psi_o) , \qquad \langle h_i^2 \rangle = (\bar{\psi}_o, h^2\psi_o)$$

$$\langle h_i h_{i+1} \rangle = (\bar{\psi}_o, hTh\psi_o)/t_o , \tag{46}$$

where $\bar{\psi}_o = e^{-\beta w}\psi_o$ is the corresponding eigenfunction of T^+.

We can readily estimate the nearest neighbor height correlation. Suppose $\bar{h} \equiv \langle h \rangle$, then

$$\chi \equiv \langle (h_i - \bar{h})(h_{i+1} - \bar{h}) \rangle / \bar{h}^2 = (\bar{\psi}_o, (h - \bar{h})T(h - \bar{h})\psi_o)/t_o\bar{h}^2 \tag{47}$$

But if the eigenvalues t_α of T are ordered from t_o down to zero (since $e^{-\lambda|h-h'|}$ is positive definite, so is $\bar{T}$), then

$$0 \leq (\bar{\psi}_o, (h - \bar{h})T(h - \bar{h})\psi_o) = \sum_0^\infty (\bar{\psi}_o, (h - \bar{h})\psi_\alpha) t_\alpha (\bar{\psi}_\alpha, (h - \bar{h})\psi_o)$$

$$= \sum_1^\infty (\bar{\psi}_o, (h - \bar{h})\psi_\alpha) t_\alpha (\bar{\psi}_o, (h - \bar{h})\psi_o) \leq t_1 \sum_1^\infty (\bar{\psi}_o, (h - \bar{h})\psi_\alpha)(\bar{\psi}_\alpha, (h - \bar{h})\psi_o)$$

$$= t_1(\bar{\psi}_o, (h - \bar{h})^2\psi_o) , \quad \text{and so}$$

$$0 \leq \chi \leq \frac{t_1}{t_o} \frac{\langle h^2 - \bar{h}^2 \rangle}{\bar{h}^2} . \tag{48}$$

To be more explicit, since $(\lambda^2 - d^2/dh^2)e^{-\lambda|h-h'|} = 2\lambda\delta(h-h')$, (44) can be rewritten

$$T^{-1} = \frac{1}{2\lambda} e^{\beta w(h)}\left(-\frac{d^2}{dh^2} + \lambda^2\right) . \tag{49}$$

Make a change of variable given by

$$dz/dh = \gamma\, e^{-\beta(w(h)-w_o)} \tag{50}$$

where $w_o = w_{min}$, $= w(\bar{h})$, as we will see; choose $z = 0$ at $h = \bar{h}$. Now (49) transforms to

$$T^{-1} = \frac{\gamma^2}{2\lambda}\, e^{\beta w_o}\left[- \frac{d}{dz}\, e^{-\beta(w-w_o)}\, \frac{d}{dz} + (\frac{\lambda}{\gamma})^2\, e^{\beta(w-w_o)}\right]. \tag{51}$$

We approximate

$$d^{-\beta(w-w_o)} = 1 - z^2, \tag{52}$$

which will be correct in the neighborhood of the minimum if

$$\gamma^2 = \frac{1}{2}\frac{\partial^2 \beta w(\bar{h})}{\partial \bar{h}^2} = \beta A(\sigma+\tau)/\bar{h}^3 \tag{53}$$

and then

$$z = \tanh\gamma(h-\bar{h}), \qquad e^{-\beta(w-w_o)} = \text{sech}^2\gamma(h-\bar{h}). \tag{54}$$

$$T^{-1} = \frac{\gamma^2}{2\lambda}\, e^{\beta w_o}\left[- \frac{d}{dz}(1-z^2)\frac{d}{dz} + \frac{m^2}{1-z^2}\right]$$

where $m = \lambda/\gamma$. The eigenvalues of this associated Legendre equation are $\ell(\ell+1)$, where $\ell = m, m+1, \ldots$, leading to

$$t_o = \frac{2\lambda}{\gamma^2}\, e^{-\beta w_o} / \frac{\lambda}{\gamma}\left(\frac{\lambda}{\gamma} + 1\right)$$

$$t_1/t_o = \frac{\lambda}{\lambda+2\gamma}, \ldots \tag{55}$$

(and $\psi_o = \text{sech}^{\lambda/\gamma}\gamma(h-\bar{h}), \ldots$). All properties can now be extracted, but we confine ourselves to the observation that as "temperature" increases, or β decreases, $t_1/t_o \propto \beta^{\frac{1}{2}}$ continually decreases the nearest neighbor correlation in comparison with the single height variance.

11. CONCLUDING REMARKS.   In this paper we have emphasized the
modeling of the mechanical response of cell aggregates to a distributed field of stress determined by cell parameters.   This is
a small part of the problem of morphogenetic movement, and one
where the models probably will conform most closely to classical
fluid mechanics.   The challenge before us is to find descriptions
of cell movements which conform to observation of embryological
events, and these must involve a range of activity not usually
present in continuum mechanics.   We have not dealt, for example,
with cell division and growth, although in many instances this is
an essential feature of movement.   In continuum theories this can
be introduced rather simply as a source of material, but in cell-
based descriptions we must contemplate the continual enlargement
of the configuration space as new cells appear.   In the two-
dimensional monolayer mentioned in Section 8, this might be
handled by "splitting" nucleii, but in that model no allowance is
made for cell polarity and orientation of the cleavage plane.

Perhaps most compelling is the need to deal with global rearrangements of cells, involving the breaking and forming of
cell-cell bonds.   At the Lagrangian level, continuum mechanics
can deal with local cell deformation via the local Jacobian matrix
(generalizing the function $dx/da = 1/h$ in the monolayer model),
and invariants of this matrix presumably play a key role in defining an acceptable effective "pressure" in a continuum limit.   But
rearrangements of cells cannot be treated in that way, and so
the cell dimension determines a length scale of discrete activity
where Lagrangian labels are discontinuous, and where fluctuations
in cell movement must be resolved.   Ultimately, therefore, the
challenge is to model the mechanical activity of individual cells.

Added note on a paper of A. Gierer: After this paper was pre-
pared the work of Gierer [17] was brought to our attention. In
his paper evagination is studied within a framework conceptually
similar to the models introduced in Section 4 above, with equilibrium structures determined at a minimum of a potential. Global
stability of aggregates of cells is studied, and applications are
given to deformation of rotationally symmetric as well as
two-dimensional cell sheets.

BIBLIOGRAPHY

1. Jacobson, A.G. and Gordon, R., "Changes in the shape of the
   developing vertebrate nervous system analyzed experimentally,
   mathematically, and by computer simulation," Jour. Exp. Zool.,
   197 (1976), 191-246.

2. Odell, G., Oster, G., Burnside, B., and Alberch, P., "A Mechanical model of epithelial morphogenesis," J. Math. Biology,
   9 (1980), 291-95.

3. Childress, S. and Percus, J.K., Mathematical models in developmental biology, CIMS Lecture Notes, New York University
   (1978).

4. Holtfreter, J., "A study of the mechanics of gastrulation,"
   J. Exp. Biol. 94 (1943), 261-318. Part II: Jour. Exp. Zool.
   95 (1944), 171-212.

5. Moscona, A., "Cell suspensions from organ rudiments of chick
   embryos," Exp. Cell Res. 3 (1952), 535-9.

6. Steinberg, M., "Reconstruction of tissues by dissociated
   cells," Science 141 (1963), 401-8. See also Mathematical
   Models of Cell Rearrangement, edited by G. D. Mostow, Yale
   University Press, New Haven, 1975.

7. Bell, G.I., "Models for the specific adhesion of cells to
   cells," Science 200 (1978), 618-27.

8. See papers in the collection edited by Mostow (loc. cit.)
   especially Goel et al., "Self-sorting of isotropic cells,"
   J. Theor. Biol., 28 (1970), 423-68.

9. Goel, N.S. and Rogers, G., "Computer simulation of engulfment
   and other movements of embryonic tissues," J. Theor. Biol.,
   71 (1978), 103-140.

10. Greenspan, H.P., "On the dynamics of cell cleavage," J. Theor.
    Biol. 65 (1970), 79-99.

BIBLIOGRAPHY, *continued*

11. Greenspan, H.P., "On the motion of a small viscous droplet that wets a surface," J. Fluid Mech. 84 (1978), 125-143.

12. Gordon, R. *et al.*, "A rheological mechanism sufficient to explain the kinetics of cell sorting," J. Theor. Biol. 37 (1972), 43-73.

13. Peskin, C., "A Lagrangian method for the Navier-Stokes equations with large deformations," preprint.

14. Caller, H.B., in Fluctuations, Relaxation, and Resonance in Magnetic Systems, edited by D. Ter Haar, Oliver and Boyd. London, 1962.

15. Gnedenko, B.V., Theory of Probability, Chelsea, New York, 1962

16. Huang, K., Statistical Mechanics, John Wiley, New York, 1963.

17. Gierer, A., "Physical aspects of tissue evagination and biological form," Q. Rev. Biophys. 10 (1977), 529-93.

COURANT INSTITUTE OF MATHEMATICAL SCIENCES
NEW YORK UNIVERSITY
NEW YORK, N.Y.   10012

Lectures on Mathematics in the Life Sciences
Volume 14, 1981

FEEDING AT LOW REYNOLDS NUMBER BY COPEPODS

M. A. R. Koehl[1]

ABSTRACT. Calanoid copepods are small planktonic
crustacens that are extremely abundant in oceans and
lakes. Many of these shrimp-like animals eat unicellular algae and thus play a major role in the transfer
of energy through marine food chains. In spite of the
ecological importance of copepod feeding, the mechanisms
by which these animals capture particles such as algal
cells has been poorly understood. Analysis of high-
speed movies of feeding copepods has revealed how
these tiny creatures move water to capture food.
Because copepods are small, their physical world is
dominated by viscous forces rather than the inertial
forces that large organisms encounter when moving
through fluids. In the viscous world of a copepod, water
flow is laminar, bristled appendages behave as solid
paddles rather than open rakes, particles can neither be
scooped up or left behind because appendages have thick
layers of water stuck to them, and water and particle
movements stop immediately when an animal stops moving its
appendages. This study of copepod feeding illustrates
the importance of considering the physical forces that
are most important at the size scale of the organisms
being studied. A number of unsolved problems about
copepod feeding that are ripe for mathematical analysis
are presented.

1980 Mathematics Subject Classification.   76Z10.
[1]This work was supported by grants from the National
Science and Engineering Research Council of Canada,
National Science Foundation Grant #OCE76-01142, and a
Faculty Research Grant, University of California.

INTRODUCTION

Why Study Copepod Feeding?

Calanoid copepods (Fig. 1) are planktonic crustaceans that are extremely abundant in oceans and lakes. These small (usually a few millimeters long) shrimp-like animals are a very important link in marine food chains (see Russell-Hunter, 1970; Cushing, 1975). Many copepods feed on unicellular plants such as diatoms and dinoflagellates. These copepods are in turn eaten by carnivorous zooplankton and small fish. Thus copepod feeding is ecologically important in a number of ways. Copepods can markedly influence not only the abundance, but also the size- and species-composition, of the phytoplankton by grazing on some species of these small plants more heavily than on others (e.g. Porter, 1977; Richman et al., 1977; McCauley and Briand, 1979). Conversely, the abundance and composition of the phytoplankton can have important effects on the growth or

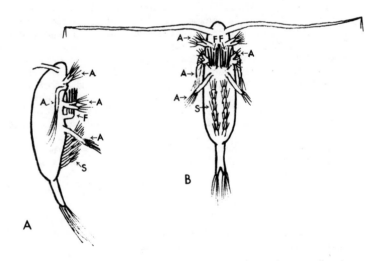

Fig. 1. Side (A) and ventral (B) views of a calanoid copepod. The filters are indicated by "f", the feeding appendages by "a" and the swimming legs by "s".

decline of populations of various species of copepods (e.g.
Mullin and Brooks, 1970; Harris and Paffenhöffer, 1976; Daag,
1977; Checkley, 1980) which in turn affect the animals of higher
links in the food chain. Furthermore, copepod feeding is of
ecological importance because copepods remove from the water not
only phytoplankton, but also particulate matter such as
detritus, fecal pellets, and spilled oil (e.g. Paffenhöfer and
Strickland, 1970; Conover, 1971).

The Copepod Feeding Controversy

In spite of the tremendous ecological importance of
copepod feeding, the mechanisms by which these animals capture
particles are poorly understood, due in part to the technical
difficulties involved in observing feeding appendages only
fractions of a millimeter long that are moving at rates of 20
to 80 Hz, and due in part to the non-intuitive nature of
viscous water flow around small objects.

Until recently, information about copepod feeding came
mainly from microscope observations of feeding currents produced by copepods in drops of water and from laboratory experiments in which the rates at which copepods removed food
particles from volumes of water was determined. A copepod was
thought to flap four pairs of legs ("feeding appendages", Fig.
1) to create a current of water, part of which was shoved continuously through the bristles on another pair of legs
("filters", Figs. 1 and 2) that were held stationary over the
mouth. Copepods tend to graze on large particles more heavily
than small ones (e.g. Marshall, 1973; Frost, 1977); it has been
suggested that such "size-selective" feeding is due to the
spacing of barbs ("setules") on the bristles ("setae") of the
filters, which act like passive sieves (e.g. Nival and Nival,
1976; Boyd, 1976). However, copepods show a plasticity of

selective feeding behavior that is difficult to explain if they
simply sieve particles out of the water (e.g. Poulet and
Marsot, 1978; Cowles, 1979; Donaghay and Small, 1979; Richman
et al., 1980). Therefore, a controversy exists in the literature as to whether copepod selective feeding is due to the
physical properties of the animals' sieve-like filters, or
rather is due to active choice by the animal (see, for example,
"The Copepod Filter-Feeding Controversy" in Kerfoot, 1980).

A Plea to Theoreticians

In this symposium on mathematical questions in biology, I
would like as a biologist to pose a number of questions about
copepod feeding to mathematicians. I will first describe the
new picture we are now working out about how copepods feed, and
I will then mention some of the physical constraints on feeding
by such small animals. I will point out a number of problems
about copepod feeding that need theoretical work. My hope is
that some mathematicians will find these problems amusing
enough to tackle.

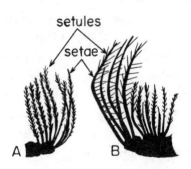

Fig. 2. Sketches of filters from two different species of
copepods that can feed on algal particles.

KINEMATICS OF COPEPOD FEEDING

It has recently become possible to make high-speed (500 frames·s^{-1}) close-up (resolution of 5 μm) movies of copepods in relatively large vessels (120 ml) of water, where artefacts in flow patterns due to confining the animals in small drops of water can be minimized (Alcaraz et al., 1980; Koehl and Strickler, 1981). A copepod in a large vessel can be kept in the field of view of the microscope for filming by holding the animal on a "leash" (i.e. gluing the animal to a fine hair that can be positioned by a micromanipulator). Such films of feeding cope-pods show the complexity of the appendage movements that create water currents that carry food towards the filters. The films also reveal that the filters are not always held stationary, but rather periodically actively capture parcels of water containing food particles.

To study the water motion produced by the feeding appen-dages of one species of copepod, Eucalanus pileatus, we marked water near feeding animals with dye released from a micropipette (Koehl and Strickler, 1981). The kinematics of copepod flapping and water movement was worked out by frame-by-frame analysis of high-speed movies of the appendage and dye stream positions. The sequence of events during one cycle of flapping is dia-grammed in Fig. 3. Note that water is not pumped through the filters when they are held still. Rather, the flapping of the four pairs of feeding appendages produces a stream of water that moves past the copepod. Low-magnification high-speed movies of untethered copepods showed that they move upwards (anteriorly) at velocities of about 1.5mm·s^{-1} when they flap their feeding appendages.

Our movies of dye streams also revealed the water motion produced by particle-capturing movements of the filters (Fig. 4). When an alga is carried into the vicinity of the

copepod, the feeding appendages beat assymetrically, redirecting
the incoming current so as to draw in water preferentially from
the direction of the alga. (If the copepod were not on a leash,
this assymetrical flapping would turn the animal towards the
alga.) As the alga nears the filters, they fling apart in a
manner similar to the vortex-creating "fling" of insect wings
(Weis-Fogh, 1973). This fling creates a gap between the
filters that is filled by inrushing water (Fig. 4,A and B).
This water carries the alga within the basket formed by the
filters, which then rapidly close over the alga and water.
While the filters are closing, the water (having no other
escape route) is squeezed out between the setae of these appendages. Water does not escape out the front of the filters as
they close because certain of the feeding legs located in front
of the filters are pushing rearward while the filters close.
Captured particles are scraped off the filters and shoved into

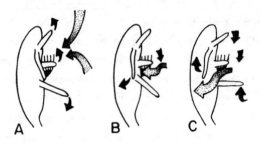

Fig. 3. Diagrams of feeding appendage movements of a copepod
(black arrows) and the water currents (stippled arrows) they
produce as revealed by high-speed movies of dye streams around
feeding Eucalanus pileatus. An arrow with a narrow shaft and
wide head indicates lateral movement out of the plane of the
page towards the reader; an arrow with a wide shaft and narrow
head indicates medial movement away from the reader. The
filter is shown in black. (A) Outward movements of the
indicated appendages suck water towards the copepod's filter.
(B) Postero-medial movements and dorsal-lateral movement of the
indicated appendages suck water laterally. (C) Inward movements
of three pairs of appendages coupled with dorso-lateral movement
of the fourth pair shove water postero-laterally.

the mouth by special comb-like structures on one of the pairs of feeding appendages.

Thus, high-speed films of water movement near feeding copepods reveal that these important herbivores propel water past themselves by flapping their feeding appendages ("scanning"), and actively capture small parcels of that water that contain food particles by flinging and closing their filters. The films also reveal that copepods stop scanning from time to time to go through an elaborate procedure of cleaning their feeding appendages, and that copepods regularly stop moving all their appendages and sink at velocities of 1 to 2 $mm \cdot s^{-1}$ (Koehl and Strickler, 1981).

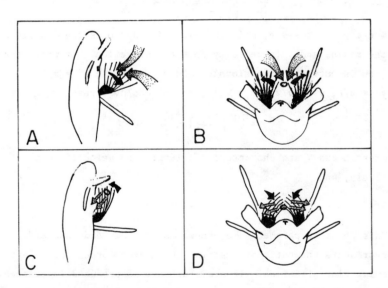

Fig. 4. Diagrams of the filter movements (black arrows) of a copepod and the water currents (stippled arrows) they produce. The filters are shown in black; the pair of feeding appendages just anterior to the filters have been omitted for clarity. The circle indicates the position of a particle being captured. The animal is viewed from its right side in A and C, and from its head end in B and D. The alga is captured by an outward fling (A and B) and inward sweep (C and D) of the filter, as described in the text.

COPEPODS LIVE IN A STICKY WORLD

In order to analyze the feeding of copepods, one must consider what the physical world is like for an organism that
small. What sorts of forces are most important to a copepod
trying to flap its feeding appendages?

A copepod's appendages are solid objects moving through
water. Consider the basic Navier-Stokes equation describing
the forces (F) affecting the motion of a Newtonian viscous
fluid,

$$F_{pressure} + F_{viscous} = F_{inertial},$$

or

$$-\frac{\partial p}{\partial x} + \mu \frac{\partial^2 v}{\partial y^2} = \rho \frac{\partial v}{\partial t} + v \frac{\partial v}{\partial x},$$

where p is pressure, μ is the viscosity and ρ is the density of
the fluid, and v is velocity in the x direction. If this equation is put into dimensionless form by letting $x' = x/L$,
$y' = y/L$, $v' = v/V$, $t' = Vt/L$, and $p' = p/\rho V^2$, then

$$-\frac{\partial p'}{\partial x'} + \frac{1}{Re}\left(\frac{\partial^2 v'}{\partial y'^2}\right) = \frac{\partial v'}{\partial t'} + v' \frac{\partial v'}{\partial x'},$$

where L and V are characteristic length and velocity respectively, and

$$Re = \frac{\rho VL}{\mu}.$$

This physical parameter Re, known as the "Reynolds number",
represents the ratio of inertial to viscous forces for a particular flow situation. If a disturbance is produced in a stream
of fluid, it will tend to persist if inertial forces predomin-
ate. Thus, for organisms operating at high Reynolds numbers
(i.e. for large, rapidly-moving organisms) the world is a
turbulent place. In contrast, at low Reynolds numbers any disturbance in the fluid tends to be damped out by the viscous
resistance of the fluid to undergoing shear deformation. Hence,

flow around small, slowly-moving organisms tends to be laminar
(i.e. the fluid movies smoothly around the body and can be considered as moving in layers between which there is no significant
mixing.)

Are copepod feeding appendages and filters small enough to
be operating at low Reynolds numbers?

The Reynolds numbers calculated for maximum velocities
attained by distal setae on various feeding appendages and on
the filter during its fling range between 10^{-2} and 10^{-1} for
E. pileatus (Koehl and Strickler, 1981). Even these maximum
Reynolds numbers are very low, indicating that inertial forces
are relatively unimportant to these copepods when they are
feeding. To get a feeling for what the world might be like for
a feeding copepod, imagine trying to remove crumbs from olive
oil using forks moving no faster than half a millimeter per
second -- also a flow situation with a Reynolds number of 10^{-2}.
The constraints that such viscous flow put on copepod feeding
are not intuitively obvious to us high Reynolds number humans.

A number of features of low Reynolds number flow, which
have been considered quantitatively (e.g. Happel and Brenner,
1965; White, 1974; Weinbaum, this symposium), and which have
been described qualitatively for copepods (Koehl and Strickler,
1981), should be kept in mind when copepod feeding is analyzed.

Laminar Flow

In the viscous, low Reynolds number world of a copepod,
water flow is laminar. By repositioning our dye-releasing
micropipette with respect to tethered copepods, we have shown
that water streams from different locations are moved around
the copepod along different discrete smooth paths. The dye is
not mixed into the surrounding water by beating copepod appendages as it would be in a turbulent, high Reynolds number flow

situation. One likely consequence of such laminar flow is that
a copepod's flapping should not mix together the water around it
and thus should not confuse the direction from which chemical
signals in the water are coming.

Water Flow Around Setae

Fluid in contact with the surface of an object does not
slip relative to that object. Thus, a layer of fluid along the
surface of a body undergoes shear deformation when the body
moves relative to the surrounding fluid. At low Reynolds number
this boundary layer of fluid surrounding the object and subject
to shear deformation is thick relative to the dimensions of the
object. Furthermore, at low Reynolds numbers when inertial
effects can be ignored, the resistance to the motion of water
between two objects depends upon the rate at which the water is
deformed in shear; the closer together the objects are, the
greater the shear deformation rate (and therefore the resistance)
will be of water forced to move between them at a given flow
rate. Thus, although copepod appendages with their long setae
(Fig. 1) look like open rakes, perhaps they behave more like
solid paddles through which water does not flow.

A rough estimate of whether or not water is likely to flow
through the gaps between setae of a copepod appendage can be
made by comparing half the distance between two neighboring
setae (s) with the thickness of the boundary layer that would
form around a solitary seta. Ellington (1975) has used this
approach for bristled insect wings. The thickness of the
boundary layer (δ) around a cylindrical seta can be estimated by

$$\delta = 0 \left(\frac{d}{(Re)^{1/2}} \right)$$

where d is the diameter and Re the Reynolds number of the
cylinder. When copepod feeding appendages flap to create the

scanning currents, $\delta > s$ (Koehl and Strickler, 1981). This is
consistent with our observations that little water moves through
the gaps between setae on these appendages (Fig. 5). Similarly,
$\delta > s$ for the setae of the stationary filters when the feeding
appendages flap and produce a water stream over the filters.
Remember that dye is observed to bypass rather than flow through
the stationary filters (Fig. 3). However, when the filters
actively capture algae, δ is the same as or slightly lower than
s (Koehl and Strickler, 1981). The capturing motions of the
filters are more rapid (δ becomes thinner as velocity is increased) and the s's are greater than they are for the setae
of the feeding appendages.

Of course water can be forced to move through the narrow
gaps between setae if given no other escape route. For example,
when the setae of the filters rapidly close over a parcel of
water that they have actively captured, water is observed to be
squeezed out between the setae (Fig. 4, C and D). Since the
closely-spaced setae of the filters should offer a great deal of
resistance to flow, such a motion might be metabolically costly.
Therefore, it makes sense for an animal only to force water
through the filters when an algal cell is there, as they do.

Fig. 5. Diagram of a feeding appendage moving towards the reader
as indicated by the arrow. Note that the black dye stream does
not flow between the setae.

PROBLEM: What is the cost of driving water through the
filters of a copepod? How does this vary with changes in the
morphology and kinematics of the filters? The metabolic cost
(energy per time) for an animal to drive water through its
filters should be a function of the resistance those filters
offer to the flow of water. Resistance is the pressure drop
(Δp) across a filter for a given flow rate (Q, volume per time)
of water through the filter. Various theoretical and empirical
approaches have been used to determine the resistance of filters
composed of cylindrical fibers to the flow of fluid through them
(for example, see reviews by Fuchs, 1964; Pich, 1966; Davies,
1973). The problem is more complicated in copepods because:
a) the tips of the setae (fibers) of the filters move more
rapidly than do their attached ends, b) the gaps between setae
are greater at their distal than at their proximal ends, and
c) the gaps between setae change with time (they become smaller
as the setae close over the captured water).

The structure of copepod filters varies from species to
species (Fig. 2) and also changes as animals grow and mature.
How do changes in the length, diameter, and spacing of setae
affect costs of driving water through the filters?

Members of different species of copepods move their filters
at different velocities. For example, filter setae of Eucalanus
pileatus close over captured water at maximum velocities of
about 20 mm·s^{-1} (Re = 8 x 10^{-2}), whereas those of Centropages
typicus do so at about 300 mm·s^{-1} (Re = 1) (Koehl and Strickler
(1981). For which copepods is this motion more expensive? It
should be pointed out that for filters operating at low Reynolds
numbers in purely viscous flow, Δp/Q is constant, whereas for
filters operating at intermediate Reynolds numbers (of the order
of 10^{-1} to 10^{1}) where flow is laminar but inertial effects
cannot be ignored, Δp/Q rises as Q rises (Davies, 1973).

How do the setules (barbs) on the filter setae (Fig. 2) affect the resistance of the filter? As a first approximation, can they be considered as simply increasing the effective diameter of the setae?

Water Flow Around Setules

Biologists have thought that water flows between the setules on the setae of copepod filters and that particles larger than the gap between the setules are sieved out of the water. However, water no doubt resists flowing between these closely-spaced setules on setae. The setae of copepod filters, covered with rows of setules and the water stuck to them, may well be functionally wide and smooth rather than comb-like. Rees (1975) has found that the corrugated wings of insects operating at low Reynolds numbers are functionally thick and smooth in this way.

PROBLEM: What is the flow field like at low Reynolds numbers around a cylinder (a seta) with smaller cylinders (setules) sticking out of it? How is the flow field affected by: a) the position and number of the setules relative to the flow direction (Fig. 6); b) the diameter and length of the setules relative to the diameter of a seta; c) the spacing of setules along the length of a seta; and d) the proximity of other similar setae bearing setules?

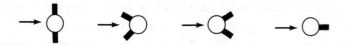

Fig. 6. Cross-section of a seta (O) with setules (I). The arrow indicates flow direction. Some examples of possible setule arrangements are shown.

The "Sphere of Influence" of a Food Particle

We have observed copepods to flap assymetrically the appendages that create the scanning current, thereby drawing in water preferentially from the direction from which an alga is arriving (i.e. turning towards the alga). We have also observed copepods to fling apart their filters when an alga is a distance of a few hundred micrometers from them. Do copepods use mechanical or chemical cues to perceive food particles?

PROBLEM: Can a flapping copepod feel distortions in the flow field it is producing caused by the presence of sinking or swimming food particles? Morphological studies suggest that copepods have mechanoreceptors (Strickler, 1975b; Strickler and Bal, 1973), and feeding experiments reveal that copepods can capture inert particles (e.g. Wilson, 1973; Poulet, 1976). Furthermore, behavioral studies indicate that various small planktonic animals feel the presence of walls and other zooplankton in the water around them (Lillelund and Lasker, 1971; Strickler, 1975b; Kerfoot et al., 1980; Zaret, 1980). At low Reynolds numbers objects affect the movement of fluid many diameters away from themselves (e.g. Happel and Brenner, 1965; White, 1974; Weinbaum, this symposium). How would the flow field produced by a copepod (which can be described empirically) be distorted by particles of a) different sizes moving at b) different velocities at c) different distances from the animal?

PROBLEM: Can copepods smell nearby food particles? The algal cells and other small particles on which copepods feed swim or sink (Eppley et al., 1967) slowly through the water at Reynolds numbers of the order of 10^{-5} to 10^{-3}. Therefore these particles are no doubt surrounded by relatively thick boundary layers of water. If the food particles exude chemicals into the water around them, it is likely that they are surrounded by

spheres of odor much larger than themselves. What is the distribution of concentrations of a chemical leaked by and diffusing away from an algal cell as it sinks or swims through the
water? How is the shape of such a field of odor distorted as
it moves through the flow field created by a copepod?

Morphological studies indicate that copepods have chemoreceptors (e.g. Fleminger, 1973; Friedman and Strickler, 1975).
Chemosensation has been demonstrated in other crustaceans (e.g.
Ache, 1972; Hamner and Hamner, 1977) and does appear to be involved in copepod mate-finding (e.g. Griffiths and Frost, 1976;
Blades and Youngblath, 1980). Some feeding experiments indicate
that copepods can preferentially feed on certain particles on
the basis of their smell (e.g. Mullin, 1963; Poulet and Marsot,
1978), although it has not yet been demonstrated whether such
selective feeding is due only to rejection of certain particles
after they have been captured, or is due also to preferential
capture of particular food.

Do scanning currents increase the amount of chemical information a copepod can receive from its environment? If diffusion
of molecules is considered in the x direction only, the rate of
change of concentration of molecules with time at a fixed distance from the source of the molecules is

$$\frac{\partial c}{\partial t} = D \frac{\partial^2 c}{\partial x^2} \ ,$$

where c is the concentration of the molecules and D is the
diffusion coefficient (D is generally of the order of $10^{-5} m^2 \cdot s^{-1}$
for molecules diffusing in water). Thus the time required for
a molecule to diffuse a distance (a) through water is roughly
$a^2/D = a^2 \times 10^5$, whereas the time to transport a molecule that
distance by moving the water in which it sits is roughly a/v,
where v is the velocity at which the water is moving (Purcell,
1977). Thus, at the velocities of scanning currents created by

flapping copepods (about 10 mm·s^{-1} in the vicinity of the
animal (Koehl and Strickler, 1981)), flapping should allow them
to smell things at distances of 0.1 µm or greater sooner than
they would if they held still and waited for molecules to diffuse to them. For example, it takes about 100x longer for a
molecule to diffuse to a copepod that is holding still from
a distance of 100 µm as it does for that molecule to be trans-
ported to the animal when it is flapping.

Can a flapping copepod receive chemical information about
the location of food particles in the water around it? Recall
that streamlines around a scanning copepod are not mixed
together. For a copepod creating a scanning current of
10 mm·s^{-1}, I estimate that molecules might diffuse only about
4 µm out of a streamline while the water moved a distance of
200 µm past the animal. Thus, chemoreceptors on the appendages
of a copepod closer to that particular streamline might well
receive more molecules from a food particle in that streamline
than would chemoreceptors on the other side of the animal.
These crude estimates indicate that it might be possible for a
copepod to receive chemical information about the location of a
food particle, but a more rigorous analysis is needed.

Theoretical approaches have been used to analyze chemoreception by organisms such as bacteria (Berg and Purcell, 1977)
and moths (Murray, 1978). Mathematical analyses could also shed
light on questions of copepod chemoreception.

Producing Water and Particle Movement at Low Reynolds Number

Since an appendage on a copepod operating at low Reynolds
number influences a thick layer of water around itself, particles move away when the appendage moves towards them (Fig. 7,
A). Thus, a copepod appendage cannot strain a particle out of
the water as we might catch a ball using a scoop net. Copepods,

rather, must maneuver particles by moving the water surrounding
the particles, as they do during the capture "fling" of the
filters (Fig. 4, A and B). Furthermore, when moving at low
Reynolds number, it is difficult to leave water behind. For
example, a copepod appendage moving away from an alga drags the
alga along (Fig. 7, B).

Since inertial effects are small at low Reynolds number,
when a copepod stops flapping its feeding appendages, the flow
around it stops almost immediately. For example, dye spots
carried in scanning currents "coasted" only 40 to 50 μm to a
halt within about 30 ms of the time copepods stopped flapping
their feeding appendages (Koehl and Strickler, 1981). At very
low Reynolds numbers, when inertia can be ignored and when
things don't coast, an organism that simply flapped its appendages back and forth symmetrically with a fast "power stroke"
and a slow "recovery stroke" would move water back and forth
along the same path rather than pushing it in some net direction
(Purcell, 1977). How do copepods, whose feeding currents only
coast very slightly, overcome this near reversibility of flow in

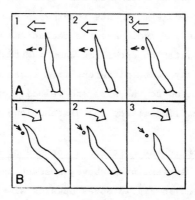

Fig. 7. Tracings of frames of a film of a copepod appendage and
an alga. The time interval between pictures is 6 ms. A. An
alga being "pushed" by the appendage. B. An alga "following"
the appendage.

space and time and propel water past themselves in some net
direction? Copepod feeding appendages follow complex, assymetrical paths such as figure-eights when they flap. Furthermore,
appendages can change their shape during a cycle of movement.
For example, the setae on some appendages are more flexible in
certain directions than in others so that at particular points
during a flap they are straightened and spread out, whereas at
others they are collapsed and bent over. In addition to these
complexities of behavior of individual appendages, the various
pairs of feeding appendages flap in different planes. Although
these appendages flap at the same frequency, they flap out of
phase with each other, thereby sucking water in and then pushing
it out from between them (Fig. 3). Clearly the qualitative
description given above of how copepods produce scanning currents should be replaced by a quantitative analysis of the fluid
dynamics of this process.

   PROBLEM: What are the mechanisms by which the movements of
copepod feeding appendages produce the water flow patterns we
observe? Is circulation set up around a copepod appendage as
it is around the wing of a tiny insect (Weis-Fogh, 1973)? How
do the feeding appendage movements together produce the net
scanning current we observe? Are the mechanisms used to propel
water by larger, faster copepods for whom inertial effects are
more important likely to be different from those used by
smaller copepods? How do the size and velocity of the filters
during the capture fling affect the distance over which they
capture water?

   PROBLEM: What is the cost for a copepod of creating the
scanning current? As mentioned above, a scanning copepod in the
orientation depicted in Fig. 3 moves upwards slowly; when it
stops scanning, it sinks slowly. Scanning is therefore somewhat
analogous to hovering flight. Perhaps the theoretical

approaches that have been used to analyze the flow around and
power requirements of a small hovering insect (e.g. Ellington,
1978) could be usefully applied to a scanning copepod.

PARTICLE CAPTURE BY THE FILTERS

Although there has been considerable speculation in the
literature about the processes by which the selective feeding
of copepods occurs, the actual mechanisms involved have not
been demonstrated. Based on our new understanding of how
copepods feed, it can be suggested that selective feeding could
depend upon the chemical or mechanical cues for which a copepod
flaps assymetrically or flings its filters, as discussed above.
Copepod selective feeding could also depend upon the physical
characteristics of particles retained within the basket formed
by the filters as they close over a parcel of captured water.
Correlations have been noted between the types of food selectively eaten by various species of copepods and the morphology
of their filters (e.g. Itoh, 1970; Boyd, 1976; Nival and Nival,
1976; Richman et al., 1980); it has therefore been suggested
that copepods capture only particles bigger than the gaps
between setules on the setae of their filters (e.g. Boyd, 1976).

Filters are Not Just Sieves

The physical mechanisms by which filter feeding organisms
remove particles from the surrounding water are poorly understood. Biologists generally assume that filters act as sieves
that only capture particles larger than the spaces between
neighboring fibers composing the filter. Rubenstein and Koehl
(1977) have applied to biological filters the theoretical
analysis of filtration developed by engineers (for reviews, see
Fuchs, 1964; Dorman, 1966; Pich, 1966; Davies, 1973); we have
suggested that there are several mechanisms other than sieving

by which a filter can capture particles, and that particles
smaller than the interfiber spacing of a filter can be caught.
Both man-made screens (Sheldon and Sutcliffe, 1969) and copepod
filters (Friedman, 1980) have been reported to retain particles
from natural waters that are smaller that the "pore size" of
the filters. Furthermore, experimental studies indicate that
this filtration theory is applicable to other particle-capturing
organisms such as brittle stars (LaBarbera, 1978), sea anemones
(Koehl, unpubl. data), and protozoans (Fenchel, 1980). Murray
(1978) has used a similar analysis to work out the capture of
pheromone molecules by moth antennae, and a number of other
investigators have used this approach to predict the deposition
of particles in the human respiratory tract (e.g. Taulbee and
Yu, 1975; Savilonis and Lee, 1977).

   PROBLEM: How do the morphology and kinematics of copepod
filters affect the types of particles they selectively capture?
The mechanisms by which filters capture particles are described
by Rubenstein and Koehl (1977) and are illustrated in Fig. 8:
sieving, direct interception, inertial impaction, gravitational
sedimentation, and motile-particle or diffusion deposition.
One can predict using certain physical characteristics of a
filter, particles, and fluid flow, which of these mechanisms of
particle capture are operative for a given filtration situation
(see Ranz and Wong, 1952; Pich, 1966).

   One of the consequences of the ability of a filter to capture particles by a number of mechanisms is that the filter
differentially captures particles of different sizes. As
particle size is increased, a filtering element's ability to
collect particles by inertial impaction, gravitational sedimentation, and direct interception is improved. As particle size
is reduced, collection by diffusion deposition is enhanced. As
a result, there is an intermediate range of particle sizes for

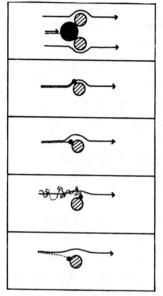

sieving

direct interception: $N_I = d_p/d_f$

inertial impaction:
$$N_I = [(\rho_m)d_p^2 V_o] /18\mu d_f$$

diffusion or motile-particle deposition:
$$N_M = \frac{KT}{d_p} \frac{1}{3\pi\mu V_o d_f}$$

gravitational sedimentation: $N_G = v_g/V_o$,
$$\text{where } v_g = [d_p^2 g(\rho_p - \rho_m)] /18\mu$$

Fig. 8. Diagrams of the mechanisms of particle capture by a fiber and the dimensionless indices (N's) indicating the intensity of particle capture by each mechanism. (Reprinted from D. I. Rubenstein and M. A. R. Koehl, Amer. Natur., 1977, Vol.111, pp. 981-984, by permission of the University of Chicago Press. © 1977 by the University of Chicago. All rights reserved.)

d_f = diameter of a fiber
d_p = diameter of a particle
g = acceleration due to gravity
K = Boltzmann's constant
m = mass of particle
N_G = index of gravitational deposition
N_I = index of inertial impaction
N_M = index of motile-particle or diffusion deposition
N_R = index of direct interception
T = absolute temperature
V_o = upstream velocity of the fluid relative to the fiber
v_g = settling velocity of the particle
μ = viscosity of the fluid
ρ_m = density of the fluid
ρ_p = density of the particle

⊘ = crossection of fiber
● = particle
⟶ = streamline of fluid relative to fiber
┄➤ = path of particle

which the efficiency of capture by the simultaneous action of
all mechanisms reaches a minimum. ("Filtering efficiency" is
defined as the ratio of the number of particles striking a
fiber in a filter to the number that would strike it if the
flow streamlines were not diverted by it (Dorman, 1966).) The
efficiency minimum occurs at smaller particle sizes as velocity
is increased and as fiber diameter is decreased (Pich, 1966).
Therefore, physical features of a filter other than just its
inter-fiber spacing determine the size range of particles the
filter is most efficient at capturing.

As mentioned above, the morphology (see Fig. 2) and kinematics of copepod filters can vary considerably from species to
species of copepod. A model of copepod filters that would allow
particle selectivity to be predicted from the structure and
movement of such filters would be extremely useful to biologists.

PROBLEM: How can a copepod change the size of particles
on which it selectively feeds? Feeding experiments indicate
that copepods can alter the size range of particles that they
preferentially capture (e.g. Cowles, 1979). If a copepod's
second maxillae are considered only as sieves, it appears that
the animal can change the size of particles on which it feeds
only by changing the spacing between the fibers of these
filters. Copepod setules are fixed structures whose spacing
on setae cannot be actively altered (and which may well be
hidden in the boundary layer around setae). Filtration theory
indicates several other means by which a copepod might alter
its diet. For example, by changing the velocity of water
passing through the filters, or by altering the diameter or
adhesiveness (with mucus?) of the setae, the range of particle
sizes that can be captured most efficiently by the filter can
be shifted. Price and Paffenhöfer (pers. comm.) have noted
that copepods move their filters differently when feeding on

very small (<7 μm) algae than they do when feeding on larger
algae. It would be useful to be able to predict how such
differences in movement would affect the size range of particles
most efficiently captured by a copepod's filters.

PROBLEM: How do the morphology and movement of copepod
filters affect their effective filter area? The area of a
filter through which water actually passes should decrease as
the variability of the spacing between fibers in the filter
increases (because water tends to flow through the wider spaces
and to avoid the narrower ones). The spacing between and
velocity of copepod setae both increase along the length of the
setae from base to tip, as does the length and density of
setules on the setae. How do these features interact to affect
the volume flow rates of water through different areas of
filters of various species of copepods?

OPTIMAL FORAGING MODELS FOR COPEPODS

Several optimal foraging models have been proposed to
predict the size-selective feeding behavior of copepods (e.g.
Lam and Frost, 1976; Lehman, 1976). Those models are based on
the assumption that copepods are simple on-off filter feeders
(i.e. when they flap their feeding appendages, they continuously
drive water through their sieve-like filters). Now that we
know the complex repertoire of behaviors used by feeding
copepods, new foraging models should be developed based on more
realistic assumptions. Copepods create scanning currents
("search"), fling and close their filters ("capture"), comb particles from the filters into the mouth ("handle"), and cease
flapping and thus sink ("rest"). Copepods can also rapidly
locomote through the water be flapping their swimming legs
(Fig. 1) (Vlyman, 1970; Strickler, 1975a; Lehman, 1977). Both
sinking and swimming may be involved in predator avoidance and

both behaviors may also move copepods to new food environments.
If reasonable estimates of the power requirements (energetic
costs per time) of these various activities could be made using
fluid dynamic analyses, and if the ways in which copepods
apportion their time between these activities under different
food conditions could be assessed (as Cowles and Strickler
(pers. comm.) and Price and Paffenhöfer (pers. comm.) are now
doing empirically), then more realistic foraging models could be
developed. Before the complexity of copepod behavior had been
directly observed, Haurey and Weihs (1976) used such an
approach to model how a copepod should apportion its time between sinking and swimming to maintain a position in the water
column at minimum cost.

CONCLUSIONS

Although copepod feeding is extremely important ecologically, the mechanisms by which it occurs are poorly understood.
Controversies rage in the literature about how copepods feed
selectively. Now that we can use high-speed microcinematography
to work out the kinematics of copepod feeding, and now that we
are aware of the non-intuitive nature of low Reynolds number
water flow around these tiny animals, we can pose a number of
questions about copepod feeding mechanisms that require mathematical analyses. This field of biological research would
certainly be advanced at this stage by the contributions of
theoreticians.

ACKNOWLEDGEMENTS

I am grateful to T. J. Cowles, G.-A. Paffenhöfer, D. I.
Rubenstein, and J. R. Strickler for their collaboration on much
of the research that led to the ideas expressed in this paper.

BIBLIOGRAPHY

Ache, B. W.   1972.   Amino acid receptors in the antennules of
    Homarus americanus.   Comp. Biochem. Physiol. 42A: 807–811.

Alcaraz, M., G.-A. Paffenhöfer and J. R. Strickler. 1980.
    Catching the algae: A first account of visual observa-
    tions on filter-feeding calanoids. p. 241–248, In W. C.
    Kerfoot [ed.] , Evolution and ecology of zooplankton com-
    munities. Univ. Press of New England.

Berg, H. C. and E. M. Purcell. 1977. Physics of chemorecep-
    tion. Biophys. J. 20: 193–219.

Blades, P. I. and M. J. Yongblath. 1980. Morphological,
    physiological, and behavioral aspects of mating in calanoid
    copepods. p. 39–51. In W. C. Kerfoot [ed.] , Evolution
    and ecology of zooplankton communities. Univ. Press of
    New England.

Boyd, C. M. 1976. Selection of particle sizes by filter-
    feeding copepods: A plea for reason. Limnol. Oceanogr.
    21: 175–180.

Checkley, D. M. 1980. Food limitation of egg production by a
    marine planktonic copepod in the sea off southern
    California. Limnol. Oceanogr. 25: 991–998.

Conover, R. J. 1971. Some relations between zooplankton and
    bunker C oil in Chedabacto Bay following the wreck of the
    tanker ARROW. J. Fish. Res. Bd. Can. 28: 1327–1330.

Cushing, D. H. 1975. Marine ecology and fisheries. Cambridge
    Univ. Press.

Cowles, T. J. 1979. The feeding responses of copepods from the
    Peru upwelling system: Food size selection. J. Mar. Res.
    37: 601–622.

Daag, M. 1977. Some effects of patchy food environments on
    copepods. Limnol. Oceanogr. 22: 99–107.

Davies, C. N. 1973. Air filtration. Academic Press.

Donaghay, P. L. and L. F. Small. 1979. Food selection capa-
    bilities of the estuarine copepod Acartia clausii. Mar.
    Biol. 52: 137–146.

Dorman, R. G. 1966. Filtration. p. 195–222. In C. N. Davies
    [ed.] , Aerosol science. Academic Press.

Ellington, C. P. 1975. Non-steady-state aerodynamics of the
    flight of Encarsia formoso. p. 783–796. In T. Y.-T. Wu
    et al. [eds.] , Swimming and flying in nature, V. 2.
    Plenum.

Ellington, C. P. 1978. The aerodynamics of normal hovering flight: Three approaches. p. 327–345. In K. Schmidt-Nielson et al. [eds.] , comparative physiology: Water, ions, and fluid mechanics. Cambridge Univ. Press.

Fenchel, T. 1980. Relation between particle size selection and clearance in suspension-feeding ciliates. Limnol. Oceanogr. 25: 733–738.

Friedman, M. M. 1980. Comparative morphology and functional significance of copepod receptors and oral structures. pp. 185–197. In W. C. Kerfoot [ed.] , Evolution and ecology of zooplankton communities. Univ. Press of New England.

Friedman, M. M. and J. R. Strickler. 1975. Chemoreceptors and feeding in calanoid copepods (Arthropoda: Crustacea). Proc. Nat. Acad. Sci. U. S. A. 72: 4185–4188.

Frost, B. W. 1977. Feeding behavior of Calanus pacificus in mixtures of food particles. Limnol. Oceanogr. 22: 472–491.

Fuchs, N. A. 1964. The mechanics of aerosols. Pergamon.

Griffiths, A. M. and B. W. Frost. 1976. Chemical communication in the marine planktonic copepods Calanus pacificus and Pseudocalanus sp. Crustacean 30: 1–8.

Hamner, P. and W. M. Hamner. 1977. Chemosensory tracking of scent trails by the planktonic shrimp Acetes sibogae australis. Science 195: 886–888.

Happel, J. and H. Brenner. 1965. Low Reynolds number hydro-dynamics. Prentice-Hall.

Haurey, L. and D. Weihs. 1976. Energetically efficient swim-ming behavior of negatively buoyant zooplankton. Limnol. Oceanogr. 21: 797–803.

Harris, R. F. and G.-A. Paffenhöfer. 1976b. The effect of food concentration on cumulative injestion and growth efficiency of two small marine copepods. J. Mar. Biol. Ass. U. K. 56: 865–888.

Itoh, K. 1970. A consideration on feeding habits of planktonic copepods in relation to the structure of their oral parts. Bull. Plankt. Soc. Jpn. 17: 1–10.

Kerfoot, W. C. [ed.] . 1980. Evolution and ecology of zoo-plankton communities. Univ. Press of New England.

Kerfoot, W. C., D. L. Kellogg, Jr., and J. R. Strickler. 1980. Visual observations of live zooplankton: Evasion, escape, and chemical defenses. pp. 10–27, In W. C. Kerfoot [ed.] , Evolution and ecology of zooplankton communities. Univ. Press of New England.

Koehl, M. A. R. and J. R. Strickler. 1981. Copepod feeding
    currents; Food capture at low Reynolds number. Limnol.
    Oceanogr. (in press).

LaBarbera, M. 1978. Particle capture by a Pacific brittle
    star: Experimental test of aerosol suspension feeding
    model. Science 201: 1147-1149.

Lam, R. K. and B. W. Frost. 1976. Model of copepod filtering
    response to changes in size and concentration of food.
    Limnol. Oceanogr. 21: 490-500.

Lehman, J. T. 1976. The filter feeder as an optimal forager,
    and the predicted shapes of feeding curves. Limnol.
    Oceanogr. 21: 501-516.

Lehman, J. T. 1977. On calculating drag characteristics for
    decelerating zooplankton. Limnol. Oceanogr. 22: 170-172.

Lillelund, K. and R. Lasker. 1971. Laboratory studies of pre-
    dation by marine copepods on fish larvae. Fishery Bull.
    69: 655-667.

McCauley, E. and F. Briand. 1979. Zooplankton grazing and
    phytoplankton species richness: Field tests of the preda-
    tion hypothesis. Limnol. Oceanogr. 24: 243-252.

Marshall, S. M. 1973. Respiration and feeding in copepods.
    Adv. Mar. Biol. 11: 57-120.

Mullin, M. M. 1963. Some factors affecting the feeding of
    marine copepods of the genus Calanus. Limnol. Oceanogr.
    8: 239-250.

Mullin, M. M. and E. R. Brooks. 1970. Growth and metabolism
    of two planktonic marine copepods as influenced by
    temperature and type of food. p. 74-95, In J. H. Steele
    [ed.] , Marine food chains. Oliver and Boyd.

Murray, J. D. 1978. Reduction of dimensionability in diffu-
    sion processes: Antenna receptors of moths. pp. 83-127,
    In Nonlinear differential equation models in biology.
    Oxford Univ. Press.

Nival, P. and S. Nival. 1976. Particle retention efficiencies
    of an herbivorous copepod, Acartia clausi (adult and
    copepodite stages): Effects on grazing. Limnol. Oceanogr.
    21: 24-38.

Paffenhöfer, G.-A. and J. D. H. Strickland. 1970. A note on
    the feeding of Calanus helgolandicus on detritus. Mar.
    Biol. 5: 97-99.

Pich, J. 1966. Theory of aerosol filtration by fibrous and
    membrane filters. p. 223-285, In C. N. Davies [ed.] ,
    Aerosol science. Academic Press.

Porter, K. G. 1977. The plant-animal interface in freshwater
    ecosystems. Am. Sci. 65: 159-170.

Poulet, S. A. 1976. Feeding of Pseudocalanus minutus on living
    and non-living particles. Mar. Biol. 34: 117-125.

Poulet, S. A. and P. Marsot. 1978. Chemosensory grazing by
    marine calanoid copepods (Arthropoda: Crustacea). Science
    200: 1403-1405.

Purcell, E. M. 1977. Life at Low Reynolds number. Am. J.
    Physics 45: 3-11.

Ranz, W. E. and J. B. Wong. 1952. Impaction of dust and smoke
    particles on surface and body collectors. Ind. Eng. Chem.
    44: 1371-1381.

Rees, C. J. C. 1975. Aerodynamics properties of an insect
    wing section and a smooth aerofoil compared. Nature
    258: 141-142.

Richman, S., S. A. Bohon, and S. E. Robbins. 1980. Grazing
    interactions among freshwater calanoid copepods. p. 219-
    233, In W. C. Kerfoot [ed.] , Evolution and ecology of
    zooplankton communities. Univ. Press of New England.

Richman, S., D. R. Heinle and R. Huff. 1977. Grazing by adult
    estuarine calanoid copepods of the Chesapeake Bay. Mar.
    Biol. 42: 69-84.

Rubenstein, D. I. and M. A. R. Koehl. 1977. The mechanisms of
    filter feeding: Some theoretical considerations. Am. Nat.
    111: 981-994.

Russell-Hunter, W. D. 1967. Aquatic Productivity. MacMillan
    Co.

Savilonis, B. J. and J. S. Lee. 1977. Model for aerosol
    impaction in the lung airways. J. Biomechanics 10: 413-
    417.

Sheldon, R. W. and W. H. Sutcliffe. 1969. Retention of marine
    particles by screens and filters. Limnol. Oceanogr.
    14: 441-444.

Strickler, J. R. 1975a. Swimming of planktonic Cyclops
    (copepoda, Crustacea): Pattern, movements and their
    control. p. 599-616, In T. Y-T. Wu et al. [eds.] , Swim-
    ming and flying in nature, V. 2. Plenum.

Strickler, J. R. 1975b. Intra- and inter-specific information
    flow among planktonic copepods: Receptors. Verh. Inter.
    Ver. Limnol. 19: 2951-58.

Strickler, J. R. and Bal, A. K. 1973. Setae of the first antennae of the copepod Cyclops scutifer (Sars): Their structure and importance. Proc. Nat. Acad. Sci. 70: 2656-59.

Taulbee, D. B. and C. P. Yu. 1975. A theory of aerosol deposition in the human respiratory tract. J. Appl. Physiol. 38: 77-85.

Vlyman, W. J. 1970. Energetic expenditure by swimming copepods. Limnol. Oceanogr. 15: 348-356.

Weinbaum, S. 1981. Particle motion through pores and near boundaries in biological flows. (this symposium).

Weis-Fogh, T. 1973. Quick estimates of flight fitness in hovering animals, including novel mechanisms for lift production. J. Exp. Biol. 59: 169-230.

White, F. M. 1974. Viscous fluid flow. McGraw Hill.

Wilson, D. S. 1973. Food selection among copepods. Ecology 54: 909-914.

Zaret, R. E. 1980. The animal and its viscous environment. pp. 3-9. In W. C. Kerfoot [ed.] , Evolution and ecology of zooplankton communities. Univ. Press of New England.

DEPARTMENT OF ZOOLOGY
UNIVERSITY OF CALIFORNIA
BERKELEY, CALIFORNIA 94720

Lectures on Mathematics in the Life Sciences
Volume 14, 1981

STRONG INTERACTION THEORY FOR
PARTICLE MOTION THROUGH PORES AND NEAR BOUNDARIES IN
BIOLOGICAL FLOWS AT LOW REYNOLDS NUMBER

Sheldon Weinbaum[1]

ABSTRACT.   Particle and boundary interactions at low
Reynolds number have traditionally been treated using
weak interaction method of reflection techniques.
This talk will describe recently developed analytical
and numerical methods for treating spheroidal particles
in strong hydrodynamic interaction with each other
and with neighboring boundaries. The theory will
first be illustrated for the hydrodynamic three body
problem and then applied to several problems of
biological interest such as (i) a hydrodynamic mechanism for the formation of red cell rouleaux,(ii) a simplified model for the transendothelial diffusion of
plasmalemma vesicles,(iii) a theory for determining the
phenomenological coefficients for the Kedem-Katchalsky
membrane transport equations and (iv) the motion of
neutrally bouyant particles at the entrance to biological pores and channels or the mouth of a foraging
micro-organism.

1. INTRODUCTION.

Much of the world of cellular level fluid mechanics
involves the strong hydrodynamic interaction between particles or
between particles and boundaries at extremely low Reynolds numbers
of the order of 10^{-3} or less.Typical examples are the interaction
between nearly identical red cells in the formation of red cell
rouleaux, the swimming and foraging of micro-organisms,the motion

---

1980 Mathematics Subject Classification.92A05,76.35,76.65.
[1]This talk describes joint research with my colleague
R.Pfeffer and several of our former graduate students
M.Gluckman, S.Leichtberg, P.Ganatos and Z.Dagan.

of the cellular components of blood through the microcirculation,
the intracellular migration of vesicles and the filtration and
diffusion of molecules across biological and artificial membranes.
One particularly intriguing phenomenon which has not been satisfactorily explained in microcirculatory fluid mechanics is the
Fahraeus effect. The hematocrit or concentration of red cells in
microvessels or fine glass tubes is found to be significantly
less than in the feed reservoir that supplies them, whereas the
exudate collected at the tube exit has exactly the same concentration as the feed reservoir.

Much of the existing theory for low Reynolds number flow
has been based on a weak interaction theory commonly called the
method of reflections, or for large aspect ratio bodies a slender
body analysis in which low order singularities are distributed
along the midchord of the body (see Happel and Brenner[2] for an
excellent summary). In the method of reflections each boundary
in the flow is first treated in the absence of the others and
then the no slip boundary conditions are satisfied by an iterative
series generated by the lowest order solution. Higher order
terms are obtained by requiring that the reflection of each other
disturbance on all other boundaries in the flow field be reduced
to zero. The mathematical difficulty with this technique and
with creeping motion flow in general is that the fundamental solution for the velocity decays as $1/r$ for three dimensional objects
with the result that an iterative series solution converges very
slowly if at all unless the boundaries are very far apart.
Similarly, end effects on slender objects and the influence of
sharp corners can spread to large distances. Thus the first nine
terms in the iterative series solution, Faxen[3], using the reflection technique yields drag results that deviate by more than 25

---

[1]Supported by the National Science Foundation Under Grant
ENG 78 - 22101

percent from the exact solution for coaxial flow past two neighboring spheres and the error in drag due to end effects on a slender body in creeping motion is approximated by $\varepsilon = (0.69 + 2.3$
$\log L/a)$,Batchelor[4], where L/a is the aspect ratio; thus a 15 per
cent error in drag can be expected for a slender body with L/a =
10^3. Another important consequence of the algebraic decay of
disturbances generated by each boundary is observed in the theory
of suspensions where the summation of particle pair interactions
is not bounded unless special techniques, Batchelor[5], are used.
An important simplification for the biological applications
considered herein is that the Oseen[6] inertial corrections in the
far field, which become important as $\frac{r}{a}$ Re approaches unity, lie
far outside the range where the strong interaction effects considered herein are important.

It is only in the past ten years that strong interaction techniques have been devised for treating more varied boundary value
problems in low Reynolds number flow. Prior to this exact solutions to the creeping motion equations involving more than one
boundary were either based on the Stimson and Jeffrey[7] solution
for axisymmetric flow past two equal spheres or the spherical
bipolar coordinate expansion introduced by O'Neill[8] and applied
by Wacholder and Sather[9] for two unequal spheres and by Brenner
and coworkers[10,11] for the limiting case of a sphere and a planar
surface. The one important exception is the series truncation
solution of Haberman and Sayre[12] for the axial motion of a sphere
along a circular cylinder. The last analysis is a precursor to
the strong interaction techniques described in the present paper.
Various investigators have also attempted to solve the Navier-
Stokes[6] equations by finite difference numerical methods in the
creeping motion regime. These efforts have encountered fundamental difficulties because of the slow asymptotic decay of the
disturbances and the singularity in the pressure field that
occurs whenever sharp edges are present. Thus, it is much easier
to obtain a numerical Navier-Stokes solution at a Reynolds number

of ten than a Reynolds number of zero.

The focus of the present talk will be on the strong interaction between spheroids or between spheroids and boundaries. More general particle shapes can be treated by integral equation techniques in which the body is represented by a surface distribution of fundamental singularities, as shown by Gluckman et al[13] for arbitrary axisymmetric shapes and generalized to fully three dimensional objects in Youngren and Acrivos[14]. The basic concepts behind the solution procedure will first be illustrated by examining the behavior of three or more closely spaced spheroids in an unbounded fluid. The complications of axisymmetric and three dimensional boundaries will then be introduced. The applications have been chosen for their biological intrest. These include: (i) the unsteady interaction between finite axial arrays of identical particles in a tube (new mechanism for rouleaux formation), (ii) a simplified model for the diffusion of plasmalemma vesicles across an endothelial cell, (iii) the three dimensional motion of a neutrally buoyant spheroid in Poiseuille channel flow, (iv) a theory to determine the filtration and diffusion coefficients in the Kedem–Katchalsky equations for membrane transport and (v) the motion of a neutrally buoyant particle at the entrance to biological pores and channels or the mouth of a foraging micro-organism.

2.   BASIC EQUATIONS.

The governing equations for the steady state creeping motion of a viscous fluid are

$$\mu \nabla^2 \bar{V} = \nabla p, \quad \nabla \cdot \bar{V} = 0, \qquad (1a,b)$$

where μ is the fluid viscosity, $\bar{V}$ the vector velocity and p the pressure. For axisymmetric flow equations (1a,b) can be combined by introducing an axisymmetric stream function

$$u = -\frac{1}{r} \frac{\partial \psi}{\partial z}, \quad w = \frac{1}{r} \frac{\partial \psi}{\partial r}, \qquad (2a,b)$$

where u and w are the radial and axial components of the fluid velocity. Taking the curl of (1a) and substituting the definition (2a,b), one obtains

$$D^2(D^2\psi) = 0 \qquad (3)$$

where D^2 is the generalized axisymmetric Stokesian operator

$$D^2 = \frac{\partial^2}{\partial r^2} - \frac{1}{r}\frac{\partial}{\partial r} + \frac{\partial^2}{\partial z^2} . \qquad (4)$$

Because the governing equations (1a,b) are linear, the general solution to (3) can be written as the superposition of three parts, the incoming flow at infinity ψ_∞, the disturbances due to the walls ψ_w and the disturbances due to the spherical particles ψ_s

$$\psi = \psi_\infty + \psi_w + \psi_s . \qquad (5)$$

The part ψ_s represents N infinite series for the $j = 1\cdots, N$ spheres containing all the simply separable solutions of (3) in spherical coordinates $(\rho_j, \theta_j, \phi_j)$ with origin at the center of each sphere which yield a vanishing fluid velocity as $\rho_j \rightarrow \infty$

$$\psi_s = \sum_{j=1}^{N} \sum_{n=2}^{\infty} (B_{n_j}\rho_j^{-n+1} + D_{n_j}\rho_j^{-n+1})I_{n_j}(\xi_j) \qquad (6)$$

Here $\xi_j = \cos\theta_j$, $I_{n_j}(\xi_j)$ is the Gegenbauer function of the first kind of order n and degree $-1/2$ and B_{n_j} and D_{n_j} are unknown constants which must be determined by satisfying all boundary conditions in the flow. The part ψ_w depends on the particular geometry of the confining walls, but will be a form of Fourier-Bessel integral for most of the applications considered herein.

In the case of three-dimensional motion, where it is not possible to introduce the simplification of a streamfunction, we construct the solution to equations (1a,b) as the superposition of three velocity disturbances

$$\bar{V} = \bar{V}_\infty + \bar{V}_w + \bar{V}_s \qquad (7)$$

describing the flow at infinity, the disturbances generated by the confining boundaries and those generated by each of the

spheres. In place of equation (6), the fundamental solutions
of equations (1a,b) describing an arbitrary velocity disturbance
emanating from the surface of N spheres with bounded behavior
as $\rho_j \to \infty$ are

$$\overline{V}_s = \sum_{j=1}^{N} \sum_{n=1}^{\infty} [\nabla \times \overline{\rho}_j X_{-(n+1)}(\rho_j,\theta_j,\phi_j) + \nabla \Phi_{-(n+1)}(\rho_j,\theta_j,\phi_j) \quad (8)$$

$$- \frac{(n-2)}{\mu 2n(2n-1)} \rho_j^2 \nabla P_{-(n+1)}(\rho_j,\theta_j,\phi_j)$$

$$\pm \frac{(n+1)}{\mu n(2n-1)} \rho_j \, P_{-(n+1)}(\rho_j,\theta_j,\phi_j)],$$

where $\overline{\rho}_j$ is a position vector measured from the center of the
j^{th} sphere and X, Φ and P are Lamb's[15] solid spherical harmonic
functions. These functions are given by

$$\begin{bmatrix} X_{-(n+1)} \\ \Phi_{-(n+1)} \\ P_{-(n+1)} \end{bmatrix} = \sum_{m=0}^{n} P_n^m(\xi_j) \frac{1}{\rho_j^{n+1}} \left\{ \begin{bmatrix} A_{jmn} \\ C_{jmn} \\ E_{jmn} \end{bmatrix} \cos.m\phi_j + \begin{bmatrix} B_{jmn} \\ D_{jmn} \\ F_{jmn} \end{bmatrix} \sin m\phi_j \right\}$$

$$(9)$$

where P_n^m is the associated Legendre function, $\xi_j = \cos\theta_j$ and A_{jmn}
$\cdots$, F_{jmn} are unknown constants to be determined by the
boundary conditions. For infinite planar confining boundaries
$\overline{V}_w$ is a double Fourier integral which will be described later.

The complicating feature of the solution representations
(5,6) or (7,8) is that there is no single orthogonal coordinate
system which can be used to satisfy the no-slip boundary conditions on the surfaces of the spheres and the confining walls
except for the special case of two spheres or a sphere and an
infinite plane. In the method of reflections this difficulty
is treated by handling each boundary in a separate sequential
manner in which the no-slip boundary conditions are satisfied on

a single coordinate surface at each step in the iterative series
solution. The fundamental premise of strong interaction theory
is that much more rapid convergence can be achieved by a truncation series solution in which all boundaries are treated
simultaneously to the same order of approximation than an iterative series generated by leading terms which are far removed
from the final solution. Truncation series arise naturally
from boundary collocation procedures in which boundary conditions
are satisfied at discrete points along a surface rather than
continuously. These procedures are particularly effective when
the individual terms in the truncation series are exact field
solutions of the governing equations and the boundary is finite.
Infinite boundaries are much more difficult to handle because
the Fourier integrals representing these boundaries are not
easily truncated especially in creeping motion where a complex
distrubance can influence the flow at large distances. Under
these circumstances it might be necessary to first satisfy the
no slip boundary conditions along all infinite confining walls
by exact analytical methods before applying the boundary
collocation procedure on the finite boundaries of the spherical
particles.

In view of the above discussion, we shall first illustrate
the boundary collocation, truncated series solution technique
for finite axisymmetric and three dimensional multiparticle
configurations in unbounded flow, and then describe the important
modifications in the basic approach required to treat confining
walls. Even without confining walls present the interaction
between three or more identical spheres is intrisically more
complicated since the particle configuration and velocity are
continuously changing due to particle interactions in contrast to
the case of two equal spheres where the spacing is fixed.

An important advantage of writing the spherical disturbances
in the form of equation (6) or (8) is that one obtains simple
expressions for the force and torque acting on each sphere. For

axisymmetric flow the force on the jth sphere is shown in Happel
and Brenner[2] to be the surface integral

$$F_j = \mu\pi\int_0^\pi \rho_j^3 \sin^3\theta_j \frac{\partial}{\partial\rho_j} [\frac{D^2\psi_j}{\rho_j^2\sin\theta_j}]\rho_j d\theta_j . \qquad (10)$$

Substituting (5) and (6) in (10), one can show using the orthogo-
nality properties of Gegenbauer functions that

$$F_j = 4\pi\mu D_{2j} . \qquad (11)$$

Similarly, for three dimensional flow it is shown in Happel and
Brenner[2] that

$$\bar{F}_j = -4\pi\nabla(\rho_j^3 P_{-2}) , \qquad (12)$$

$$\bar{T}_j = -8\pi\mu\nabla(\rho_j^3 X_{-2}) . \qquad (13)$$

If the spheres are confined to a single vertical plane (y=0),
the case considered in the next section, one finds after substituting (9) into (12) and (13), see Ganatos et al.[16],

$$\bar{F}_j = -4\pi[E_{j11} \bar{j} + E_{j01} \bar{k}] , \qquad (14)$$

$$\bar{T}_j = -8\pi\mu B_{j11} \bar{j} . \qquad (15)$$

Thus the force and torque on any sphere can be expressed in terms
of the lowest order constant coefficients of the rotational terms
in the infinite series representations (6) and (8). The value of
these constants in equation (11) or (14) and (15), however, will
depend on all the higher order constant coefficients in the series
representation for all the spheres. Results (11), (14) and (15)
are equally valid when confining walls are present, provided the
appropriate symmetry conditions are preserved.

3.   UNBOUNDED SPHERES STEADY FLOW.
        The simplest problem to consider is the coaxial uniform
flow past a fixed line array of equal spheres in an infinite fluid

This flow geometry has the additional advantage that the convergence of the boundary collocation technique can be carefully compared with the exact solution of Stimson and Jeffrey[7] for the coaxial flow past two identical spheres. The details of the solution procedure are described in Gluckman et al[17] and we shall describe only the essential features here.

From (5) and (6) the total streamfunction for this flow is given by

$$\psi = 1/2 U \rho^2 \sin^2\theta + \sum_{j=1}^{N} \sum_{n=2}^{\infty} (B_{nj}\rho_j^{-n+1} + D_{nj}\rho_j^{-n+3}) I_{nj}(\xi_j) \quad (16)$$

when ρ and θ are measured from any convenient origin along the axis. Since the no-slip boundary conditions are given on the velocity rather than the streamfunction, equation (16) must be differentiated. To do this all the ρ_j and θ_j must be expressed in terms of the same coordinate system with the same origin as chosen for ρ and θ. This procedure is algebraically lengthy but straightforward. One notes from equation (16) that if the series for each of the N spheres were truncated after M terms there would be a total of 2xNxM unknown B_{nj} and D_{nj} coefficients. Since the no-slip boundary conditions require that two velocity components vanish at each point along the generating arc of each sphere, one obtains a linear matrix of 2xNxM algebraic equations for the B_{nj} and D_{nj} coefficients by requiring that the no slip boundary conditions be satisfied exactly at M discrete points on each of the N spheres. The subroutines for the inversion of a linear system of algebraic equations are very rapid so that as many as several hundred boundary points can be treated simultaneously using only a few seconds of computer time. One minor complication discussed in Gluckman et al[17] is that the matrix is singular for boundary points both on the axis and at the uppermost point on the generating arc of the spheres. Except for these locations the choice of boundary points is not critical. A mirror image pair of points is taken within one degree of the

$\theta_j = \frac{\pi}{2}$ location (considered as one point in table 1) and the remaining points chosen at equal distances along the generating arcs of the spheres.

Table 1 is an abbreviated table taken from[17] showing the rather remarkable convergence of the series truncation solution for the drag on each sphere to the exact two sphere solution of Stimson and Jeffrey[7]. It is common practice to express the drag on each sphere in terms of a correction factor λ_j defined by

$$F_j = 4\pi\mu D_{2j} = 6\pi\mu u a \lambda_j \qquad (17)$$

which is the ratio of the drag on the jth sphere to the Stokes drag on an isolated sphere moving at the same velocity.

Number of Points, M	Spacing, $\frac{d}{a}$	λ	λ (exact)
1	1	0.66152	0.64515
3	1	0.64411	
9	1	0.64515	
3	2	0.74244	0.74226
5	2	0.74226	
3	4	0.84414	0.84412
5	4	0.84412	
3	8	0.91454	0.91454

Table 1.   Convergence to Stimson Jeffrey[7] exact solution for uniform coaxial flow past two spheres.

One concludes from table 1 that even for the worst case of two spheres touching, d/a = 1, only three boundary points (θ_j = $\pi/4$, $\pi/2$, $3\pi/4$) are required when using an axisymmetric stream-function representation for equal spheres to determine the drag on each sphere to an accuracy of 0.1 per cent. Similarly, five figure accuracy for the drag can be obtained with M = 9 for the extreme case d/a = 1.   Thus, for this simple case at least, the convergence of the boundary collocation series solution is a vast improvement over the method of reflections iterative series solution. This rapid convergence opens the possibility for treating coaxial particle arrays of up to 100 spheres using a few seconds of computer time.

One sample calculation showing the essential features of coaxial particle interaction is shown in figure 1 where the drag on each sphere in equally spaced coaxial arrays of 3 to 101 spheres in a uniform stream with $d/a = 2$ is plotted.

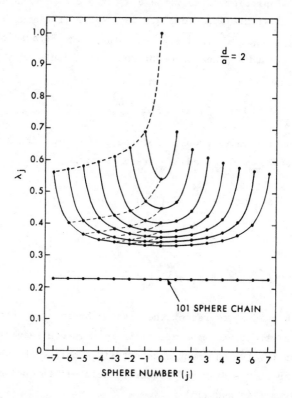

Figure 1.   Drag correction factor λ_j (shown as dots) for each sphere in equally spaced chains of spheres of increasing length with $d/a = 2$; taken from Gluckman, Pfeffer and Weinbaum[17].

The important observations in figure 1 are that the drag on the inner spheres is always less than those near the end of the chain, that the drag on any given sphere from the end of a chain decreases as the size of the chain increases and that end effects

have a very pronounced effect on even very long chains (compare
drag in 15 and 101 sphere chains) in accord with our previous
comments about the accuracy of slender body theory for low Reynolds number flow . In fact the drag per sphere in an infinite
chain approaches zero.

A simple problem to test the convergence of the three dimensional series truncation technique is the gravitational settling
of two or more spheres in a vertical plane y=0. Because of the
symmetry about this plane three of the coefficients in (9)
vanish

$$A_{jmn} = D_{jmn} = F_{jmn} = 0, \tag{18}$$

and the boundary value problem for the three remaining sets of
coefficients B_{jmn}, C_{jmn} and E_{jmn} reduces to ,

$$V_j + a\omega_j\cos\theta_j = \sum_{j=1}^{N} \bar{V}_{sj} \cdot \bar{j} ,$$

$$0 = \sum_{j=1}^{N} \bar{V}_{sj} \cdot \bar{j} , \qquad \text{on } r_j = a \tag{19}$$

$$W_j - a\omega_j\sin\theta_j\cos\phi_j = \sum_{j=1}^{N} \bar{V}_{sj} \cdot \bar{k} ,$$

where V_j, W_j and ω_j represent the unknown horizontal, vertical
and angular velocity components of the jth sphere respectively.
In addition to (19) one must also satisfy a balance between
buoyancy and Stokes drag and require a zero net torque on each
sphere.   From (14) and (15)

$$-4\pi[E_{j11}\bar{j} + E_{j01}\bar{k}] + \frac{4}{3}\pi a^3(\rho_s - \rho)\bar{k} = 0 \tag{20}$$

$$-8\pi\mu B_{j11} = 0 \tag{21}$$

To perform the differential vector operations in (8) required in
the evaluation of the velocity components on the right hand side
of (19), one must first express all the ρ_j, θ_j and ϕ_j in terms
of a single x,y,z coordinate system. This process is algebrai-

cally cumbersome but straighforward. Equations (19, (20) and (21) provide $3 \times N \times (M+1)$ linear algebraic equations for the B_{jm}, C_{jm}, E_{jm} coefficients and the unknown velocity components when the no slip boundary conditions are satisfied at M points on each of the N spheres. Owing to the planar symmetry, one can limit the choice of points to hemispherical surfaces in the region $y \geq 0$.

The solution of the linear matrix equation just described is much more sensitive to the selection of boundary collocation points than the axisymmetric streamfunction solutions previously discussed. One finds that the matrix is ill-conditioned unless $M = 2$, 4 or 12 (this is required for a uniform convergence of the coefficients) and that the optimum choice of boundary points depends on sphere orientation. These details are fully discussed in Ganatos et al[16]. The convergence of the three dimensional series truncation is compared with the exact bipolar coordinate expansion solution of Goldman et al[11] for two equal spheres falling side by side in an abbreviated table 2 taken from reference [16].

(a) Vertical drag correction factor λ_j

Spacing d/a	Exact Solution[16]	M = 2	M = 4	M = 12
1.1276	0.7327	0.7327	0.7329	0.7328
2	0.8368	0.8411	0.8369	0.8368
4	0.9135	0.9141	0.9135	0.9135
8	0.9551	0.9552	0.9551	0.9551

(b) Angular velocity $\omega_j = \omega_j a / V_t$

Spacing d/a	Exact Solution[16]	M = 2	M = 4	M = 12
1.1276	0.1314	0.9312	0.1312	0.1315
2	0.04670	0.04025	0.04676	0.04666
4	0.01172	0.01127	0.01174	0.01172
8	0.002930	0.002901	0.002932	0.002930

Table 2.   Convergence of series truncation solution to exact solution of Goldman, Cox and Brenner[11] for two spheres falling side by side. V_t terminal settling velocity isolated

sphere. Taken from Ganatos, Pfeffer and Weinbaum[16].

While the convergence of the three-dimensional truncation series
shown in table 2 is not quite as rapid as the axisymmetric solution shown in table 1, one can still calculate both the drag and
the angular velocity to an accuracy of better than 0.1 per cent
down to a fluid gap of roughly a tenth of a sphere radius using
only four boundary points on each hemisphere.

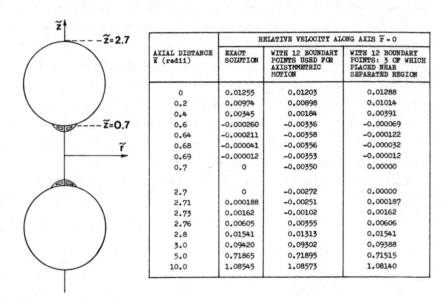

AXIAL DISTANCE $\tilde{z}$ (radii)	RELATIVE VELOCITY ALONG AXIS $\tilde{r} = 0$		
	EXACT SOLUTION	WITH 12 BOUNDARY POINTS USED FOR AXISYMMETRIC MOTION	WITH 12 BOUNDARY POINTS: 3 OF WHICH PLACED NEAR SEPARATED REGION
0	0.01255	0.01203	0.01288
0.2	0.00974	0.00898	0.01014
0.4	0.00345	0.00184	0.00391
0.6	-0.000260	-0.00336	-0.000069
0.64	-0.000211	-0.00358	-0.000122
0.68	-0.000041	-0.00356	-0.000032
0.69	-0.000012	-0.00353	-0.000012
0.7	0	-0.00350	0.00000
2.7	0	-0.00272	0.00000
2.71	0.000188	-0.00251	0.000187
2.73	0.00162	-0.00102	0.00162
2.76	0.00605	0.00355	0.00606
2.8	0.01541	0.01313	0.01541
3.0	0.09420	0.09302	0.09388
5.0	0.71865	0.71895	0.71515
10.0	1.08545	1.08573	1.08140

Figure 2.   Comparison of three dimensional boundary collocation
       series solution for local velocity field with exact solution
       of Stimson and Jeffrey[7] for coaxial flow past two equal
       spheres d/a = 1.7.   Taken from Ganatos et al [16].

The excellent convergence characteristics of the boundary
collocation series is not limited to average properties like the
drag or angular velocity but also applies to local flow behavior.
In figure 2 we show the local streamline pattern and compare the

local velocity field predicted by the three dimensional collocation method with the exact solution[7] for coaxial flow past two
equal spheres. Davis et al[18] have shown that regions of
separated flow will exist when the sphere to sphere spacing is
less than 1.79 diameters. A spacing d/a = 1.7 was therefore
chosen since small regions of recirculating flow occur on the
inner aspects of the spheres. It is evident from the results
that the 12 point collocation solution is able to reproduce such
fine detailed flow features as small separation bubbles.

In Ganatos et al[16] results are presented for the flow perpendicular to fixed horizontal arrays of equally spaced spheres. The
results for the drag are qualitatively similar to those observed
in figure 1. If the spheres are allowed to rotate about a fixed
position the angular velocity increases monotonically from zero
at the central sphere to a maximum on the outer sphere. The
maximum angular velocity is one order of magnitude smaller than
the translational velocity when d/a = 1.5 and two orders smaller
when d/a = 4.

The foregoing results indicate that muliparticle configurations are intrinsically unsteady, that particles in the central
region of a cluster will have a lower drag than those near the
edge, that the collective settling velocity of a cluster increases
significantly as the size of the cluster is increased and that
the rotation of the spheres is a higher order correction at least
down to a spacing of approximately 1.5 diameters. To study these
unsteady interactions, the importance of initial conditions and
the relative magnitude of the various unsteady terms in the
governing equations that heretofore have been omitted, we shall
briefly explore the three body problem in low Reynolds number
flow.

4.   UNSTEADY INTERACTIONS BETWEEN SPHERES.

The rapid convergence of the truncation series and the
small amount of computer time needed to invert the linear colloca-

tion matrices when a modest number of boundary points are used
make it feasible to examine the time **dependent** motion of small
sphere configurations of up to 10 spheres and to follow this
motion over a distance of the order of 10^3 sphere diameters. A
three sphere interaction in which 5000 quasi-steady 4 point
boundary collocations are performed on the surface of each sphere
requires approximately 10 minutes on an IBM 370/168 computer.

The momentum equation for unsteady creeping motion is

$$\rho \frac{\partial \bar{V}}{\partial t} = -\nabla P + \mu \nabla^2 \bar{V} .$$ (22)

The solutions to equation (22) differ from those of equation (1a)
in that they describe the time dependent diffusion of the vorti-
city boundary layer generated at the surface of an accelerating
boundary and also the induced pressure field produced by the
growth of the displacement boundary layers at these surfaces.
Following the solution procedure developed in Landau and
Lifshitz[18], one can show that the dynamic equation governing the
motion of the jth sphere is given in dimensionless form by

$$Re_\infty (\overset{\sim}{\rho} + \bar{\bar{\alpha}}_j) \frac{d \overset{\approx}{\bar{V}}}{d\tilde{t}} = 9(\bar{k} - \bar{\bar{\lambda}}_j \cdot \overset{\sim}{\bar{V}}_j) - \frac{9Re_\infty^{1/2} \bar{\bar{\beta}}_j}{\sqrt{2\pi}} \int_0^t \frac{d\overset{\approx}{\bar{V}}_j}{d\overset{\sim}{\tau}}$$

$$\frac{d\overset{\sim}{\tau}}{\sqrt{t - \overset{\sim}{\tau}}} ,$$ (23)

where $\bar{\bar{\alpha}}_j$, $\bar{\bar{\lambda}}_j$, and $\bar{\bar{\beta}}_j$, the virtual mass, viscous resistance and
Basset force tensors for sphere j, depend on the instantaneous
velocity and position of all spheres and boundaries in the flow
field and $\bar{k}$ is the dimensionless gravity vector. Dimensionless
variables denoted by a tilda are defined by

$$\overset{\approx}{\bar{V}}_j = \frac{\bar{V}_j}{V_t}, \quad \tilde{t} = \frac{tV_j}{a} , \quad \overset{\sim}{\rho} = \frac{\rho_s}{\rho} \quad \text{and} \quad Re_\infty = \frac{2aV_t}{\nu},$$

where V_t is the terminal settling velocity of an isolated sphere in a gravitational field and ρ_s is its density. An equivalent equation to (23) can also be written for the torque on each sphere. For quasi-steady settling in the limit of zero Reynolds number equation (23) reduces to:

$$\bar{k} - \bar{\bar{\lambda}}_j \cdot \bar{V}_j = 0. \tag{24}$$

In Leichtberg et al.[20] approximate numerical solutions are presented to equations (23) and (24) for the coaxial settling of three spheres in which $\bar{\bar{\alpha}}_j$ and $\bar{\bar{\beta}}_j$ are assumed to take on their single sphere values of 1/2 and unity in that order and $\bar{\bar{\lambda}}_j$ is accurately calculated using the strong interaction truncation series solution method described in the last section. Accurate solutions for $\bar{\bar{\alpha}}_j$ are presented in[20] which show that the single sphere value is a good approximation except for very close sphere spacings and does not significantly effect the integration of (23). The solutions for the velocity field show that if the spheres are released from rest there is a short initial transient

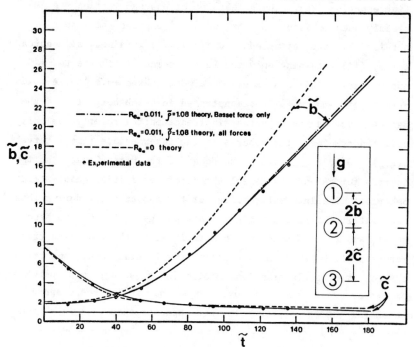

Figure 3.   Comparison between theory and experiment for change
         in sphere spacing with time for coaxial settling of three
         spheres at Re_∞ 0.011 and $\tilde{\rho}$ = 1.08. Initial spacing $\tilde{b}(0)$ =
         1.63 and $\tilde{c}(0)$ = 7.38.   ————— , all forces equation
         (23); ——— · ———   Basset only unsteady force; ––– Re_∞
         = 0 theory equation (24).

period in which the spheres rapidly approach their initial quasi-
steady state setting velocity. During the first part of this
initial transient all terms in equation (23) are of comparable
importance. Particle acceleration and virtual mass forces, however, decay exponentially on a time scale of $O(Re_\infty)$ and are of
negligible importance thereafter. The Basset force, on the other
hand, decays algebraically and produces an integrated effect in
which the particle configuration slowly drifts with increasing
time from the zero Reynolds number solution obtained by integrating equation (24). These features of unsteady motion are
clearly seen in figure 3 where we have compared the results
obtained by integrating equation (23), (i) retaining all unsteady
terms, (ii) retaining only the Basset force, (iii) the $Re_\infty = 0$
approximation (24) and experimental data taken in a low Reynolds
number settling tank. As demonstrated in Leichtberg et al[20] the
Basset force has an important influence on the relative velocity
difference between the spheres in the intermediate time interval
$O(Re_\infty) < \tilde{t} < 0(10^3 Re_\infty)$ between the decay of the virtual mass and the
inertia forces and the establishment of the initial quasi-steady
behavior. The integrated effect of the Basset force during this
time interval produces a significant change in the $\bar{V}_j$ and hence
the λ_j and is thus able to appreciably alter the long time trajectories of the spheres by changing the initial conditions for
the long time scale transient motion. The close agreement between
theory and experiment in figure 3 when the single sphere approximation for the Basset force is used suggests that this is a good
first approximation for this term when more than one sphere is
present. The difficult problem of accurately determining

$\overline{\beta}_j$ for multiparticle interactions still needs to be solved.

Another essential feature of unsteady particle interactions
is that they depend critically on initial conditions. For example, for the initial configuration shown in figure 3, sphere 2 is
initially close enough to sphere 1 to fall as a doublet and will
asymptotically approach the lead sphere 3 with a vanishing gap
width (an infinite force and time is required to completely
drain the lubricating film) and then fall with sphere 3 as a
doublet leaving sphere 1 behind. However, there is a minimum
critical initial spacing for this to occur and if this critical

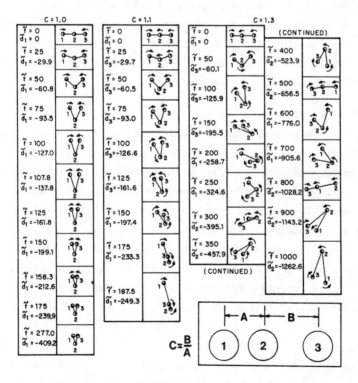

Figure 4.   Effect of initial conditions on a horizontal chain of
      three equal spheres settling freely under gravity. Initial
      spacing A + B = 6 diameters with C = B/A as indicated.

spacing is exceeded sphere 2 will asymptote to a finite fluid gap
as time goes to infinity. A more graphic example of this sensitivity to initial conditions is shown in figure 4 for different
horizonal configurations of spheres falling in a vertical plane.
Small differences in the spacing of the central sphere from the
right outer B and left outer A sphere, where C = B/A, will lead
to completely different final states for the motion after the
spheres have tumbled and changed position relative to one another
several times. In some instances such as figure 4(c) the sphere
will have to fall more than 1000 diameters before a lead doublet
is formed and the remaining sphere is left behind. Because of
the reversiblity of Stokes flow when Re_φ = 0 the trajectory of
the particles after they have formed a horizontal chain is the
mirror image of the trajectory before including the angular rotation as indicated by half darkened circles.

5.   BOUNDED SPHERES ON THE AXIS OF A TUBE.

        The motion of spheroids which are periodically and axisy-
mmetrically arranged along the axis of an infinite circular cylinder can be treated by boundary collocation techniques that resem-
ble closely those already described for flow past a coaxial chain
of spheres. The infinite tube can be broken up into periodic
cells and boundary collocation points applied along the walls of
the tube using a truncation series whose individual terms are the
separable solutions to equation (3) written in cylindical coordinates. This technique has been applied by Wang and Skalak[21] and
Hyman and Skalak[22] for an infinite periodic array of spheres and
spheroidal bubbles in a tube in a simplified model of red cells
moving through a small blood vessel. Skalak et al[23] have also
treated this same problem using an approximate finite element
method and have obtained results which were in very good agreement with the converged series truncation solutions in reference[21].
The method was then applied to irregular but identical particles
with periodic spacing to model periodic chains of identical red

cell rouleaux flowing through an arteriole or venule.

Non-periodic particle-particle and particle-wall interaction problem in tube flow are more difficult to treat because of unequal particle interactions, chain end effects and the slow axial decay of the disturbances along the tube wall. I shall outline the key steps in the solution procedure since the general method of approach is the same as that employed in the more complicated boundary value problems that follow. Starting with equations (5) and (6), one can write the total streamfunction representation for an arbitrary coaxial array of N equal spheres with unit radii in a tube of radius β with Poiseuille flow at infinity as

$$\psi = V\beta^2[\tfrac{1}{2}(\tfrac{r}{\beta})^2 - \tfrac{1}{4}(\tfrac{r}{\beta})^4] + \sum_{j=1}^{N} \sum_{n=2}^{\infty} (B_{nj}\rho_j^{-n+1} + D_{nj}\rho_j^{-n+3}) \quad (25)$$

$$+ \int_0^{\infty} [A(w)rI_1(rw) + C(w)r^2I_0(rw)]\cos wz\,dw \ .$$

The Fourier integral in (25) is an exact solution of (3) which describes an arbitrary axisymmetric disturbance along the cylinder wall which is bounded along the tube axis and vanishes as z approaches infinity. The boundary value problem is to determine the Fourier spectral functions $A(w)$ and $C(w)$ and sphere coefficients B_{nj} and D_{nj} by simultaneously satisfying the viscous flow boundary conditions along the tube wall and sphere surfaces. The principal difficulty is that $A(w)$ and $C(w)$ are complicated functions of an unknown interaction between the N spheres and the tube wall. Thus, one must first use analytical methods to satisfy the no-slip boundary conditions on the tube wall for an arbitrary coaxial N sphere disturbance, before the boundary collocation series technique can be applied on each of the spherical boundaries to determine the B_{nj} and D_{nj} coefficients.

In order to apply the no slip boundary condition on the cylinder surface $r = \beta$, one first transforms all the spherical disturbances to a common cylindrical coordinate system and

differentiates the streamfunction representation (25) to obtain
general expressions for the radial and axial velocity components.
When these velocity components are set equal to zero on the tube
wall, one obtains a set of integral equations for $A(w)$ and $C(w)$
whose form is shown below:

$$\int_0^\infty \{A(w)wI_0(w\beta) + C(w)[w\beta I_1(w\beta) + 2I_0(w\beta)]\} \cos(wz)dw$$

$$= -\sum_{j=1}^N \sum_{n=2}^\infty [B_{nj}F_n^{(1)}(z_j) + D_{nj}F_n^{(2)}(z_j)] \qquad (26)$$

$$\int_0^\infty [A(w)\beta I_1(w\beta) + C(w)\beta^2 I_0(w\beta)] \cos(wz)dw$$

$$= -\sum_{j=1}^N \sum_{n=2}^\infty [B_{nj}F_n^{(3)}(z_j) + D_{nj}F_n^{(4)}] \qquad (27)$$

The right hand sides of equations (26) and (27) describe the
spherical disturbances at locations z_j along the tube axis and
the $F_n^{(i)}$ are combinations of Legendre functions.

    The ability to solve equations (26) and (27) analytically,
or their equivalent for other axisymmetric problems, is crucial
to the success of the solution procedure. In this case one
recognizes that (26) and (27) are Fourier cosine integral
representations of their right hand sides and thus may be easily
inverted. This procedure leads to two linear algebraic equations for the $A(w)$ and $C(w)$ coefficients:

$$A(w)wI(w\beta) + C(w)[w\beta I_1(w\beta) + 2I_0(w\beta)]$$

$$= -\sum_{j=1}^N \sum_{n=2}^\infty [B_{nj}G_{nj}^{(1)}(w) + D_{nj}G_{nj}^{(2)}(w)] \qquad (28)$$

$$A(w)wI_1(w\beta) + C(w)\beta^2 I_0(w\beta) = -\sum_{j=1}^N \sum_{n=2}^\infty [B_{nj}G_{nj}^{(3)}(w)$$

$$+ D_{nj}G_{nj}^{(4)}(w)] \qquad (29)$$

where

$$G_{nj}^{(k)}(w) = \frac{1}{\pi} \int_{-\infty}^{\infty} F_n^{(k)}(z_j) \cos(wz_j) dz_j$$

The integrals representing the $G_{nj}^{(k)}$ functions are evaluated analytically and equations (28) and (29) solved for $A(w)$ and $C(w)$.

The solutions for $A(w)$ and $C(w)$ just described are now substituted back into equation (25). Since A and C are linear algebraic functions of the B_{nj} and D_{nj} spherical coefficients, equation (25) can be rewritten in the form

$$\psi = V\beta^2[\tfrac{1}{2}(\tfrac{r}{\beta})^2 - \tfrac{1}{4}(\tfrac{r}{\beta})^4] + \sum_{j=1}^{N} \sum_{n=2}^{\infty} [B_{nj}S_{nj}^{(1)}(r,z) \cdot$$

$$+ D_{nj}T_{nj}^{(1)}(r,z)], \qquad\qquad (30)$$

where the disturbances due to the spheres and the cylinder walls have been combined into a common series representation. The $S_{nj}^{(1)}$ and $T_{nj}^{(1)}$ involve integrals of w which have to be performed numerically. The solution (30) satisfies the boundary conditions at infinity and the no slip conditions on the cylinder walls independent of the values of the B_{nj} and D_{nj} coefficients. The boundary value problem for satisfying the no slip boundary conditions on the surface of the sphere has thus been reduced to the same functional form as equation (16) for a coaxial array of spheres in unbounded flow. The matrix equation for determining the B_{nj} and D_{nj} coefficients is set up using the same boundary collocation procedure as discussed in section 3. The inversion of this matrix is computationally somewhat more time consuming since it also requires the numerical evaluation of the integrals in the $S_{nj}^{(1)}$ and $T_{nj}^{(1)}$ functions. Solutions are presented in Leichtberg et al[24] for systems of up to nine spheres.

Some representative solutions for the drag correction factor on each sphere in coaxial sphere arrays of increasing length moving with velocity V in a tube are shown in figure 5 for a sphere-tube diameter ratio of 0.5. The infinite chain

results are from Wang and Skalak[21]. The effect of increasing the
length of the chain and the drag on each sphere as a function of
its position in the chain are qualitatively similar to figure 1.
The results for d/2a = 2 clearly indicate the important effect
that the wall plays in damping out particle-particle interactions
even for the case where the sphere occupies only one-quarter of
the cross sectional area of the tube.

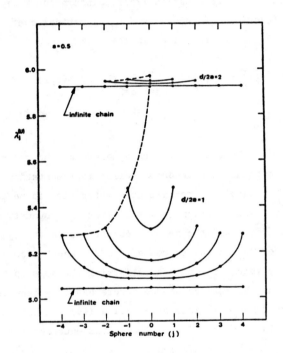

Figure 5.   Drag correction factor $\lambda_j^{(u)}$ for each sphere in
       finite coaxial chains of spheres. Sphere to tube diameter
       ratio 0.5. Taken from Leichtberg, Pfeffer & Weinbaum[24].

    A biological problem which has attracted interest is whether
identical red cells moving along the axis of a tube can aggregate
due to unequal hydrodynamic interaction. To answer this question
I shall explore the difference in axial velocity between
individual neutrally buoyant spheres in a finite coaxial array

that is carried along the tube centerline by the flow in zero
drag motion. This flow situation is the superposition of two motions, a chain of spheres moving with unequal velocity U_j in a
quiescent fluid ($V = 0$), where the drag on each sphere is given
by

$$F_j = 4\pi\mu D_{2j}^{(u)} = 6\pi\mu a U_j \lambda_j^{(u)}, \tag{31}$$

and the Poiseuille flow with centerline velocity V past a stationary array ($U_j = 0$) where

$$F_j = 4\pi\mu D_{2j}^{(v)} = -6\pi\mu a V \lambda_j^{(v)} \tag{32}$$

For zero-drag motion

$$D_{2j}^{(u)} + D_{2j}^{(v)} = \frac{3a}{2}(V_j \lambda_j^{(u)} - V\lambda_j^{(v)}) = 0$$

or

$$\frac{U_j}{V} = \frac{\lambda_j^{(v)}}{\lambda_j^{(u)}} \tag{33}$$

Since the $\lambda_j^{(u)}$ are functions of all the unknown U_j, it is not
convenient to use equation (33) directly. Instead we set the
D_{2j},which are the superposition of $D_{2j}^{(u)}$ and $D_{2j}^{(v)}$, equal to zero
in the matrix equation (30) for their combined motion and treat
the U_j as the unknowns. The number of unknowns thus remains the
same.

In figure 6 we have plotted the zero drag velocity ratio
U_j/V for each sphere in sphere chains of one to nine spheres in
a tube with a diameter ratio 0.5 for an instantaneous initial
sphere spacing $d/2a = 1.1$. The fascinating result observed in
figure 6 is that for zero drag motion the instantaneous velocities of the central spheres in the chain is slower than the end
spheres. This behavior is just oppsite that found for the coaxial settling of three or more spheres in a gravitational field or
the drag on a fixed chain of spheres in either bounded or unbounded coaxial flow. The difference in zero drag velocity, however,

is small, the maximum difference being of the order of a half of a percent for the diameter ratio shown. The results suggest that

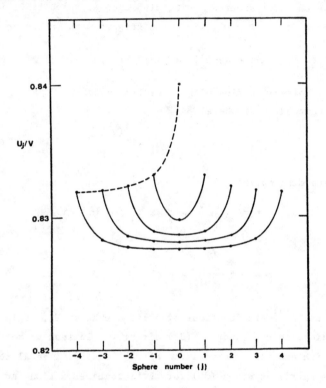

Figure 6.   Zero-drag velocities for finite chains of 1 to 9
      spheres for a sphere spacing d/2a = 1.1 and a sphere-tube
      diameter ratio 0.5. Taken from Leichtberg, Pfeffer &
      Weinbaum[24].

finite axial arrays of identical particles will slowly change
their configuration with spheres near the rear half of the chain
aggregating and those in the front half spreading apart. This
new mechanism for red cell aggregation or rouleaux formation is
described in more detailed in Leichtberg et al[25] where time
dependent calculations for the motion of neutrally buoyant chains
of spheres on the centerline of a Poiseuille flow are performed.

6.   MOTION OF A SPHERE BETWEEN PLANE PARALLEL BOUNDARIES.

The motion of a particle between plane parallel walls has important application in a variety of cellular level transport problems. I shall first describe how the three dimensional boundary collocation technique outlined in section 2 for unbounded spheres can be extended to include planar confining boundaries and then illustrate this theory by considering the gravitational settling of a sphere in an inclined channel and the motion of a neutrally buoyant sphere in a Poiseuille channel flow. In sections 7 and 8 the theory will be used to model the diffusion of a plasmalemma vesicle across an endothelial cell and the filtration and diffusion of a spherical molecule along the intercellular cleft between adjacent cells in a cell layer.

For the relatively simple axisymmetric flow configurations considered until now the viscous resistance tensor $\bar{\bar{\lambda}}_j$ could be described by a single component, the drag correction factor λ_j defined in equation (17). For even what appear to be simple three dimensional flows this is not true and one must construct the motion as a superposition of two or more boundary value problems which when summed describe the motion of interest. To illustrate the superposition consider the gravitational settling of a sphere of radius a in an inclined channel making an angle β with respect to the horizontal whose parallel walls are spaced a distance d apart. In quasi-steady low Reynolds number flow this problem can be considered as the linear superposition of three problems. If the x,y,z coordinate axes are taken such that the motion occurs in an x,z plane, where x and z are parallel and perpendicular to the walls of the channel , this superposition is composed of (i) a translation without rotation parallel to the channel walls in the x direction, (ii) a translation perpendicular to the channel walls in the z direction and (iii) a rotation without translation about the y axis, all three motions occuring in quiescent fluid. The force and torque balance to describe this motion requires five non-zero resistance coefficients F_x^t, F_x^r, F_z^t,

T_y^t, T_y^r

$$F_x = 6\pi\mu a(UF_x^t + a\Omega F_x^r) + \frac{4}{3}\pi a^3(\rho_s - \rho)g\sin\beta = 0, \quad (34)$$

$$F_z = 6\pi\mu aWF_x^t + \frac{4}{3}\pi a^3(\rho_s - \rho)g\cos\beta = 0, \quad (35)$$

$$T_y = 8\pi\mu a^2(UT_y^t + a\Omega T_y^r) = 0 \quad (36)$$

and U, W and Ω are the unknown translational and angular velocity components. F_x^t and T_y^t are the force and torque coefficients for problem (i), F_z^t the force coefficient for problem (ii) (torque coefficient is zero since problem (ii) is axisymmetric) and F_x^r and T_y^r the force and torque coefficients for problem (iii). Note that F_z^r is zero since a rotation about the y axis generates no normal force unless the sphere deforms or inertia effects are considered.

The solution of equations (34), (35) and (36) is conveniently written in terms of the dimensionless velocity ratios

$$U/U_t = T_y^r \sin\beta/(F_x^t T_y^r - F_x^r T_y^t), \quad (37)$$

$$W/U_t = -\cos\beta/F_z^t \quad , \quad (38)$$

$$a\Omega/U_t = T_y^r \sin\beta/(F_x^t T_y^r - F_x^r T_y^t), \quad (39)$$

where $U_t = (2a^2/9\mu)(\rho_s - \rho)g$ is the terminal settling velocity of an isolated sphere. From (37) and (38) the trajectory of the settling sphere is given by

$$dx/dz = U/W = F_z^t T_y^r \tan\beta/(F_x^t T_y^r - F_x^r T_y^t) = f(z) \quad (40)$$

The general problem has thus been decomposed into three simpler problems which we shall first need to solve to determine the five

resistance coefficients at any point in the flow field. We shall
see shortly that the resistance tensor is highly non-isotropic
with the result that sphere falls and rotates in a non-vertical
trajectory whose angle with respect to the gravity vector changes
as a function of the sphere's position between tha walls.

The resistance tensor for the three dimensional coplanar
settling of three or more spheres discussed earlier is also non-
isotropic, however, in that problem it was not practical to decom-
pose the tensor into its various components. In the multi-sphere
problem one would need to calculate two translational and one
rotational tensor components for each configuration of the test
sphere relative to the instantaneous positions and velocities of
all the other spheres in the system where the velocity components
are unknown. It was thus simpler to bypass the process of determining the friction tensor and solve for the unknown translational
and angular velocities directly as part of the matrix equation for
the sphere coefficients by including equations (20) and (21) in
the matrix.

Problem (ii), the perpendicular motion to the planar walls,
is axisymmetric and therefore can be treated by methods very
similar to those already outlined in the previous section for
coaxial flow of spheres in a circular tube. The Fourier-Bessel
integral representing the cylindrical boundary in equation (25) is
now replaced by

$$\psi_w = \int_0^\infty [A(w)e^{wz} + B(w)e^{-wz} + C(w)wze^{wz} + D(w)wze^{-wz}]rJ_1(wr)dw, \tag{41}$$

where r is a radial coordinate measured from a z axis passing
through the center of the sphere parallel to the planar boundaries.
Note that equation (41) represents an arbitrary axisymmetric disturbance from two infinite planar surfaces and contains four unknown spectral functions to provide the freedom to satisfy the no-
slip boundary conditions on two velocity components at each wall.

In Ganatos et al[26] $A(w)$, $B(w)$, $C(w)$, and $D(w)$ are determined analytically for any set of sphere coefficients B_{nj} and D_{nj} by in-
verting in closed form the Fourier-Bessel transform of the velocity disturbance evaluated at each of the planar surfaces. When
the solutions for the spectral functions are substituted back into
equation (41) and the spherical disturbance ψ_s added to ψ_w, one
obtains a streamfunction representation whose functional form is
equivalent to equation (30). The B_{nj} and D_{nj} coefficients are
now determined by the boundary collocation techniques previously
described.

Figure 7 shows a comparison of the boundary collocation
solutions presented in Ganatos et al[26] with the first order
method of reflection   solutions obtained by Ho and Leal[27] for
the drag correction factor for the tranverse motion of a sphere
between two plane walls. The spacing parameter s defined in the
accompanying sketch is zero for the limiting case of a sphere
moving perpendicular to a single plane wall and 0.5 for a sphere
midway between two walls. The boundary collocation solutions
shown have converged to four significant digits with the exact
spherical bipolar series solutions of Brenner[10] for the single
wall case. While it is perhaps not surprising to see that the
first order method of reflection solution is off by an order of
magnitude when the fluid gap with the closer wall is a tenth of a
radius b/a = 1.1 (one would expect weak interaction theory to be
a poor approximation at these close spacings), one observes
that the weak interaction results are more than 20 percent in
error when b/a = 5 and the channel width is five or more sphere
diameters.

Problems (i) and (iii), translation parallel to or rotation
between two plane parallel boundaries can be treated within the
framework of the more general three-dimensional flow situation
depicted in figure 8 where there is an incoming profile $V_\infty(z)$.
The solution is therefore easily specialized to include a
Poiseuille channel flow or a linear Couette flow.

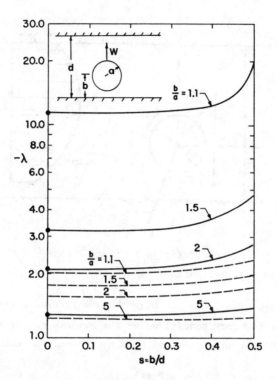

Figure 7.   Comparison between solutions for drag correction
   factor λ for a sphere moving perpendicular to plane parallel
   walls. ——— strong interaction solution, Ganatos,Weinbaum and
   Pfeffer[26]; --- method of reflections, Ho and Leal[27];• exact
   bispherical series, Brenner[10].

The form of the solution for the flow configuration in figure
8 is the superposition of the three velocity disturbances in
equation (7). The flow at infinity is

$$\bar{V}_\infty = V_\infty(z)\ \bar{j}\ .\qquad(42)$$

The velocity disturbance due to the sphere,equations (8) and (9),
can be significantly simplified because of the planar symmetry
about the y = 0 plane. Several of the coefficients in (9) are

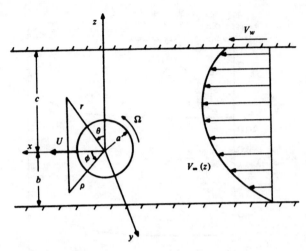

Figure 8.   Flow configuration for translation and rotation of a
    sphere in parallel channel flow.

zero (see equation (18)) and only the $m = 1$ terms are required to
satisfy the boundary conditions on the surface of the sphere.
Equation (9) for the spherical harmonic functions reduces to

$$X_{-(n+1)} = B_{1n}(\frac{1}{\rho^{n+1}})P_n'(\xi) \sin\phi,$$
$$\Phi_{-(n+1)} = C_{1n}(\frac{1}{\rho^{n+1}})P_n'(\xi) \cos\phi, \qquad (42)$$
$$P_{-(n+1)} = E_{1n}(\frac{1}{\rho^{n+1}})P_n(\xi) \cos\phi.$$

The resistance coefficients in equations (34), (35) and (36) can
be related to the lowest order constants in equation (42) using
equations (14) and (15) for the force and torque. Thus for
problems (i) and (iii), one obtains

$$F_x^t = -\frac{4}{3}\frac{B_{11}^t}{aU}, \quad T_y^t = -\frac{E_{11}^t}{a^2 U} \qquad (43)$$

$$F_x^r = -\frac{4}{3}\frac{B_{11}^r}{a^2 \Omega}, \quad T_y^r = -\frac{E_{11}^r}{a^3 \Omega}. \qquad (44)$$

Similarly, for Poiseuille flow past a rigidly held sphere between
stationary walls

$$\bar{F} = 6\pi\mu aVF_x^p \bar{j}, \quad \bar{T} = 8\pi\mu a^2VT_y^p \bar{j}, \qquad (45)$$

where the force and torque coefficients are given by

$$F_x^p = -\frac{4}{3}\frac{B_{11}^p}{aV}, \quad T_y^p = \frac{-E_{11}^p}{a^2V} . \qquad (46)$$

The velocity disturbance due both planar boundaries can be represented by a double Fourier integral

$$\begin{aligned}
\bar{V}_w = {}&_0\!\int^\infty {}_0\!\int^\infty D_1(\alpha,\beta,z,)\cos\alpha x\cos\beta y d\alpha d\beta \; \bar{i} \\
&+ {}_0\!\int^\infty {}_0\!\int^\infty D_2(\alpha,\beta,z,)\sin\alpha x\sin\beta y d\alpha d\beta \; \bar{j} \qquad (47) \\
&+ {}_0\!\int^\infty {}_0\!\int^\infty D_3(\alpha,\beta,z,)\sin\alpha x\cos\beta y d\alpha d\beta \; \bar{k}
\end{aligned}$$

where the integrand functions

$$\begin{aligned}
D_1(\alpha,\beta,z) = {}&[A_1(\alpha,\beta)(1 + \frac{\alpha^2}{k}z) - A_2(\alpha,\beta)\frac{\alpha\beta}{k}z \\
&- A_3(\alpha,\beta)\alpha z]e^{kz} + [B_1(\alpha,\beta)(1 - \frac{\alpha^2}{k}z) \\
&+ B_2(\alpha,\beta)\frac{\alpha\beta}{k}z - B_3(\alpha,\beta)\alpha z]e^{-kz}
\end{aligned}$$

$$\begin{aligned}
D_2(\alpha,\beta,z) = {}&[-A_1(\alpha,\beta)\frac{\alpha\beta}{k}z + A_2(\alpha,\beta)(1 + \frac{\beta^2}{k}z) \\
&+ A_3(\alpha,\beta)\beta z]e^{kz} + [B_1(\alpha,\beta)\frac{\alpha\beta}{k}z \\
&+ B_2(\alpha,\beta)(1 - \frac{\beta^2}{k}z) + B_3(\alpha,\beta)\beta z]e^{-kz}
\end{aligned}$$

$$\begin{aligned}
D_3(\alpha,\beta,z) = {}&[A_1(\alpha,\beta)\alpha z - A_2(\alpha,\beta)\beta z + A_3(\alpha,\beta)(1 - kz)]e^{kz} \\
&+ [B_1(\alpha,\beta)\alpha z - B_2(\alpha,\beta)\alpha z + B_3(\alpha,\beta)(1-kz)]e^{-kz}
\end{aligned}$$

$$k^2 = \alpha^2 + \beta^2$$

are fundamental solutions of equations (1a,b) in rectangular coordinates with planar symmetry about $y = 0$. Six unknown functions $A_j(\alpha,\beta)$ and $B_j(\alpha,\beta)$ $j = 1,2,3$ are required to satisfy the no slip boundary conditions for each velocity components on the two walls.

The boundary value problem for determinng the constants B_{1n}, C_{1n} and E_{1n} in the spherical series and the Fourier spectral

functions $A_j(\alpha, \beta)$ and $B_j(\alpha, \beta)$ is similar to the bounded axisymmetric flow problem already considered in that the unknown spectral
functions for the wall disturbance must be satisfied independent
of the values of the spherical constants. This requires that the
spherical disturbance, equation (8), be written in rectangular
coordinates, each velocity component in equation (7) be evaluated
at both walls and the double Fourier integral be inverted analytically after satisfying the no slip boundary conditions. This
lengthy mathematical procedure is performed in Ganatos et al[28].
One eventually obtains six linear algebraic equations for the
spectral functions A_j and B_j in terms of the B_{1n}, C_{1n} and E_{1n}.
When the solutions for the A_j and B_j functions are substituted
back in (47) and the coefficients of the spherical constants
combined, one obtains an expression for the local fluid velocity
of the form

$$\bar{V} = V_\infty(z)\bar{j} + \sum_{n=1}^{\infty} [B_{1n}F_n^1(x,y,z) + C_{1n}G_n^1(x,y,z)$$
$$+ E_{1n}H_n^1(x,y,z)]\bar{j} + \sum_{n=1}^{\infty} B_{1n}F_n^2(x,y,z) + C_{1n}G_n^2(x,y,z)$$
$$+ E_{1n}H_n^2(x,y,z)]\bar{j} + \sum_{n=1}^{\infty} B_{1n}F_n^3(x,y,z) + C_{1n}G_n^3(x,y,z)$$
$$+ E_{1n}H_n^3(x,y,z)]\bar{k}, \qquad (48)$$

where the F_n^1, G_n^1, H_n^1 etc. are complicated but known functions of
the coordinates and flow geometry.

Equation (48) is easily applied to any one of the three
flow situations described in equations (43), (44) and (46) by
choosing $V_\infty(z)$ and evaluating $\bar{V}$ at discrete points on the surface
of the sphere. The boundary collocation series can be truncated
at any even value of n. The convergence of this series is conveniently examined by looking at the limiting case of a sphere translating parallel to a single plane wall and comparing the collocation solutions for the drag and torque coefficients in equation
(43) with the bispherical infinite series solution of Goldman et
al[11]. This comparison is shown in table 3 where convergence to

fcur significant figures is obtained for spacings of up to a half
a sphere radius using only eight boundary collocation points and
up to roughly a tenth of a radius using sixteen collocation points.

	M	$\frac{b}{a}$ =1.13	$\frac{b}{a}$ =1.54	$\frac{b}{a}$ =2.35	$\frac{b}{a}$ =3.76	$\frac{b}{a}$ =10.1
F_x^t	4	-4.084	-1.598	-1.310	-1.174	-1.059
	6	-2.176	-1.567	-1.308	-1.174	-1.059
	8	-2.140	-1.567	-1.308		
	10	-2.148				
	12	-2.151				
	14	-2.151				
	EXACT	-2.151	-1.567	-1.308	-1.74	-1.059
T_y^t	4	-0.3807	0.01118	0.002478	0.0004118	8.747×10^{-6}
	6	0.08567	0.01504	0.002651	0.0004218	8.775×10^{-6}
	8	0.07882	0.01465	0.002642	0.0004216	8.775×10^{-6}
	10	0.07518	0.01465	0.002642	0.0004216	
	12	0.07398				
	14	0.07375				
	16	0.07372				
	EXACT	0.07372	0.01465	0.002642	0.0004216	8.774×10^{-6}

Table 3.   Convergence of boundary collocation series solution
for F_x^t and T_y^t for sphere translating parallel to one wall.
Taken from Ganatos, Pfeffer and Weinbaum[28].

I shall return briefly now to the problem considered at
the begining of this section, the gravitational settling of a
sphere in an inclined channel. Converged solutions for all the
force and torque coefficients in equations (43), (44) and (46)
are presented in Ganatos et al[28] and for F_z^t in Ganatos et al[26].
These results were substituted in equation (40) and the latter
integrated to obtain typical sphere trajectories. A representative solution fo a channel inclined at 45 degrees,whose wall spacing is twice the sphere diameter,is shown in figure 9. The
dashed vertical line is the imaginary vector from its initial

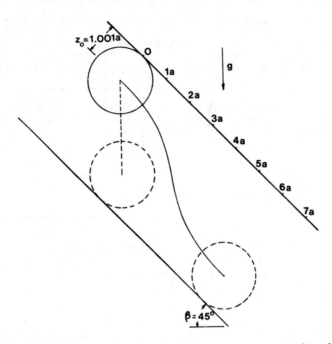

Figure 9.   Trajectory of a sphere settling in an inclined
    channel with β = 45 degrees and d/2a = 2.

position z_0. The large departure of the actual trajectory from
the imaginary path   clearly demonstrates the non-isotropic
character of the friction tensor and its dependence on sphere
position. For small fluid gaps the sphere rolls nearly parallel
to the boundary since the resistance to perpendicular motion is
much larger than to parallel motion. The motion is also axisymmetric about the midplane of the channel because of the reversibility of Stokes flow.

    A problem that has been of long standing biological interest,
because of its application to red and white cell dynamics and the
filtration flow of molecules across a cell layer, is the motion of
a neutrally buoyant particle in Poiseuille flow. This problem is
currently under study for the cylindical tube geometry.   The

channel flow geometry can be considered as the superposition of
the motions described in equations (43), (44) and (46); (i) a
parallel translation with velocity U, (ii) a rotation with angular
velocity Ω about the y axis in a quiescent fluid and (iii) a Poiseuille flow past a stationary sphere. The force and torque balance
for a neutrally buoyant sphere are:

$$F_x = 6\pi\mu a(UF_x^t + a\Omega F_x^r + VF_x^p) = 0, \qquad (49)$$

$$T_y = 8\pi\mu a^2(UT_y^t + a\Omega T_y^r + VT_y^p) = 0. \qquad (50)$$

Solving equations (49) and (50) for U and Ω, one obtains

$$\frac{U}{V} = \frac{F_x^r T_y^p - F_x^p T_y^r}{F_x^t T_y^r - F_x^r T_y^t} \qquad (51)$$

$$\frac{a\Omega}{V} = \frac{F_x^p T_y^t - F_x^t T_y^p}{F_x^t T_y^r - F_x^r T_y^t} \qquad (52)$$

Thus the sphere's translational and angular velocity can be
expressed solely in terms of the resistance coefficients defined
in equations (43), (44) and (46).

The flow quantity which is usually of greatest interest
is the local slip velocity $V_s = V(z) - V_\infty(z)$ between the sphere
and the incoming profile. The first strong interaction solutions
for this problem are shown in figure 10. One observes that the
sphere lags the Poiseuille profile everywhere and that as one
approaches the wall the slip velocity between the phases becomes
very substantial exceeding 25 percent of the local fluid velocity
for b/a < 0.1 for all physically possible values of d/2a.

7.   SIMPLIFIED MODEL FOR TRANSENDOTHELIAL DIFFUSION OF PLASMA-
     LEMMA VESICLES.

There are a number of biological applications where

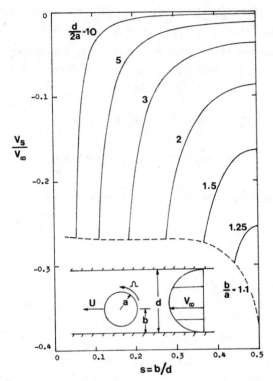

Figure 10.   Slip velocity $V_s = U - V_\infty$ of a neutrally buoyant
    sphere in Poiseuille channel flow.   Taken from Ganatos,
    Weinbaum & Pfeffer[29].

diffusion occurs in a direction normal to two nearly parallel
boundaries. One application which has recently received considerable attention is the transendothelial diffusion of plasmalemma
vesicles illustrated schematically in figure 11.

    As first suggested by Palade[30] and subsequently studied by
numerous investigators, the transendothelial transport of macro-
molecules, especially those larger than 40 Å in diameter, is
mediated primarily by 700 Å diameter intracellular membrane bound
bodies called plasmalemma vesicles. Plasmalemma vesicles differ
from endocytotic vesicles in that they are much smaller, are

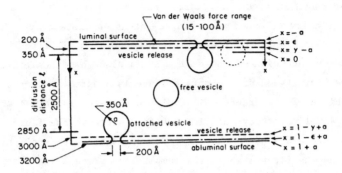

Figure 11.   Schematic diagram showing coordinates and geometry
    for plasmalemma vesicle diffusion between plasmalemma
    membranes on luminal and tissue sides of an endothelial cell.
    Dimensions shown for canine carotid arterial endothelium.

of nearly uniform size and derive their energy for motion from
Brownian collisions as opposed to a metabolically active process.
These passive vesicles are believed to form during the initial
growth of a cell and remain nearly constant in population thereafter.   A vesicle in attaching at the lumen or tissue front does
not disappear as part of the plasmalemma membrane, but forms a
quasi-stable attachment stalk which permits the vesicle to either
release its contents or fill when in the open attached configuration.   When the vesicle neck ruptures due to thermal agitation the
free vesicle can either diffuse across the cell or diffuse towards
the same membrane from which it was released.   In contrast,
endocytotic vesicles form from an active process triggered by
external stimuli in which extracellular fluid and the triggering
particle is encapsulated by an internal invagination of the
membrane.   Endocytotic transport is also directed toward the
cell's lysosomes and thus does not contribute to transcellular

transport.

As shown in figure 11 plasmalemma vesicles are assumed to be
symmetrically released at a finite intrusion distance y = 200 Å
equal to the length of the vesicle attachment stalk from the membranes on each side of the cell. It is also hypothesized that
in the vicinity of each membrane there is a narrow region ε with
dimensions of the order of 15 to 100 Å wherein attractive molecular level electrodynamic interactions overcome the fluid
resistance and lead to the fusion of the vesicle and cell membranes as the first step in the complicated process leading to the
formation of the vesicle neck. In the simplified quasi- one
dimensional steady state model described herein, I shall consider
only the vesicle diffusion in the interior of the cell and the
influence of the hydrodynamic interaction with the planar membranes on each side of the cell. We wish the model to determine
the probability that a vesicle released at the luminal front x=y
will cross the cell, hence produce a transendothelial flux ϕ_r, and
the concentration profile for vesicles released at this front.

The equation governing steady state one dimensional diffusion
with a spatially varying diffusion coefficient D(x) is

$$\frac{d}{dx}\left(D(x)\frac{dc}{dx}\right) = 0 \ , \qquad\qquad (53)$$

where c is the free vesicle concentration and

$$D(x) = \frac{kT}{6\pi\mu a\lambda(x)} = \frac{D_0}{\lambda(x)} \ . \qquad\qquad (54)$$

D_0 is the Stokes-Einstein expression for the diffusion coefficient for a diffusing particle in an infinite medium and λ(x) is
the hydrodynamic interaction function. λ is simply the resistance coefficient F_z^t introduced in equation (35) for the perpendicular motion of a spherical particle between plane parallel
walls. Equation (53) is valid in the regions ε < x < y and
y < x < ℓ-ε on each side of the release plane x = y. The matching conditions at x = y are that the concentration be continuous

and that the vesicle release rate ϕ be the sum of the diffusion
fluxes into each region,

$$c(y^-) = c(y^+) \ , \tag{55a}$$

$$\phi = D(y) \left(\frac{dc(y^-)}{dx} - \frac{dc(y^+)}{dx} \right) \ . \tag{55b}$$

The boundary conditions at $x = \varepsilon$ and $x = \ell - \varepsilon$ are that the free
vesicle concentration vanish

$$c(\varepsilon) = c(\ell - \varepsilon) = 0. \tag{56a,b}$$

A closed form solution to the boundary value problem defined
by equations (53), (55) and (56) for any $\lambda(x)$ can be obtained by
introducing the coordinate transformation

$$d\xi = \lambda(x)dx \ . \tag{57}$$

Using transformation (57), equation (53) reduces to a constant
coefficient equation for which the boundary and matching conditions are easily satisfied. This solution in dimensionless form
is

$$c(\xi) = (1 - \frac{\xi(y)}{\xi(1 - \varepsilon)})\xi \qquad \varepsilon < x < y \tag{58a}$$

$$c(\xi) = (1 - \frac{\xi}{\xi(1 - \varepsilon)})\xi(y) \qquad y < x < 1-\varepsilon \tag{58b}$$

where ξ is related to x for an arbitrary hydrodynamic resistance
law by equation (57). The solutions shown in figure 7 are a reasonable approximation for $\lambda(x)$ when vesicle-vesicle interactions
in the cell interior can be neglected and the attached vesicle
density is small. The dimensionless concentration in (58) is
defined as $CD_0/\ell\phi$ and all lengths have been scaled by ℓ. The probability ϕ_r/ϕ that a vesicle released at $x = y$ will reach and
unload its contents at the tissue front is given by

$$\phi_r/\phi = \frac{\xi(y)}{\xi(1 - \varepsilon)} \tag{59}$$

The earliest models of vesicle diffusion Tomlin[31] and Shea et al[32] neglected both the large variations in hydrodynamic resistance across the cell and the effect of the molecular force layer ε. These features were first introduced into a steady state dynamic model by Weinbaum and Caro[33]. The model was generalized for an arbitrary hydrodynamic resistance law and extended to time dependent vesicle labelling in Arminski et al[34]. The interested reader is refered to Weinbaum and Chien[35] for a survey of vesicle transport models to date.

8. A MODEL FOR DETERMINING THE PHENOMENOLOGICAL COEFFICIENTS IN THE KEDEM-KATCHALSKY MEMBRANE EQUATIONS.

The solutions obtained in section 6 for the translation and rotation of a spherical particle in parallel channel flow provide a rigorous hydrodynamic framework for calculating the phenomenological coefficients in the widely used Kedem-Katchalsky membrane equations. These equations describe the solute flux j_s and total volume flux j_v due to transmembrane pressure and concentration differences Δp and Δc for a non-electrolyte in terms of three phenomenological coefficients; ω a diffusive permeability, L_p an hydraulic permeability and σ a reflection coefficient.

$$j_s = \omega RT \, \Delta c + (1 - \sigma)\bar{c} \, j_v \, , \tag{60}$$

$$j_v = L_p(\Delta p - \sigma RT \, \Delta c) \, . \tag{61}$$

$\bar{c}$ is some average concentration in the membrane and σ takes on the values 0 or 1 for non-selective and impermeable membranes, respectively. When $\sigma = 1$ filtration flow does not contribute to the solute flux across the membrane, whereas when $\sigma = 0$ the volume flux is due solely to pressure filtration of the bulk fluid.

Most experimental evidence now indicates that the intercellular channels between adjacent cells in a cell layer are the primary ultrastructural pathway for the passage of small molecules across the cell layer though very small molecules like water and small ions can cross the cell membrane directly. The intercellular channel or cleft is very nearly a two dimensional slit with a fluid gap of roughly 200 $\overset{o}{A}$ except for localized regions of closer apposition called tight junctions where the fluid gap narrows to between 20 and 40 $\overset{o}{A}$ depending on the species and cell layer function. For present purposes I shall treat the intercellular cleft as a constant area channel of aribtrary but slowly varying cross sectional area using standard quasi one-dimensional simplifications. Models for realistic junction geometries are currently under study. These models assume that the pressure and particle concentration vary only with x and that there is an excluded layer equal to the sphere radius at each wall.

The governing equations for the motion of the solute molecule are obtained by equating the hydrodynamic force on a neutrally buoyant test particle (at these dimensions gravity can be neglected) to the gradient of the chemical potential and the net torque to zero

$$6\pi\eta a [UF_x^t + a\Omega F_x^r + V_c F_x^p] = -\frac{1}{N_a}[\frac{RT}{c}\frac{dc}{dx} + \bar{V}_s\frac{dP}{dx}], \qquad (62)$$

$$8\pi\eta a^2 [UT_y^t + a\Omega T_y^r + V_c T_y^p] = 0, \qquad (63)$$

N_a is Avogadro's number, R is the universial gas constant, T is absolute temperature, c(x) is the concentration of solute, $\bar{V}_s$ is the molar volume of solute and dc/dx and dP/dx are concentration and pressure gradients parallel to the slit walls. Solving (63) for the angular velocity Ω and substituting into (62) allows (62) to be recast into the form

$$\frac{6\pi\eta a}{F(\alpha,s)} [U - G(\alpha,s)V_c] = \frac{1}{N_a}[RT\frac{dc}{dx} + \bar{V}_s\frac{dP}{dx}], \qquad (64)$$

$$F(\alpha, s) = \frac{-T_y^r}{F_x^t \, T_y^r - F_x^r \, T_y^t} \, , \tag{65}$$

$$G(\alpha, s) = \frac{F_x^r \, T_y^p - F_x^p \, T_y^r}{F_x^t \, T_y^r - F_x^r \, T_y^t} \, , \tag{66}$$

where $\alpha = 2a/d$ is the ratio of particle size to channel width, $s = b/d$ denotes particle position and V_c is the centerline velocity of the undisturbed Poiseuille profile. The F and G functions in (65) and (66) therefore depend only on the force and torque coefficients introduced in equations (43), (44) and (46). Prior to Ganatos et al[28] there were no accurate solutions for these coefficients and a variety of ad hoc assumptions were used to estimate the F and G functions, see,e.g.,Levitt[37], Curry[38].

The functions F and G have   simple physical interpretations. If we set $V_c = 0$ in (64), one recognizes that F is the reciprocal of the drag correction factor for a spherical particle translating and rotating in an otherwise quiscent fluid when acted upon by an external force parallel to the channel walls. At the channel centerline the particle motion is a pure translation, whereas near the wall the motion is due principally to the sphere's rotation as previously seen in figure 9 for the gravitational settling of a sphere in an inclined channel. If we set $U = 0$ in (64), G/F is recognized as the drag correction factor for a sphere that is free to rotate but not translate in a Poiseuille channel flow.

The expressions for the membrane coefficients can be directly related to the F and G functions defined in (65) and (66) following the procedure used by Levitt[37]. The channel flux at any station x

$$J_s = 2\int_{a/d}^{1/2} c(x) \, U(x,s) ds = \text{const.} \tag{67}$$

when integrated along the length of the channel L yields

$$J_s = \frac{2}{L} \int_0^L \int_{a/d}^{1/2} c(x)U(x,s)ds \, dx \quad . \tag{68}$$

Solving (64) for $U(x,s)$ and substituting into (68) one finds

$$J_s = \omega[RT\nabla c + \bar{V}_s \overline{c\Delta P}] + (1 - \sigma)\bar{V}_w J_w \bar{c} \quad . \tag{69}$$

Here,

$$\bar{V}_s = \frac{4}{3}\pi a^3 N_a, \qquad \bar{c} = \frac{1}{L} \int_0^L c \, dx, \tag{70}$$

$$\Delta c = \int_0^L \frac{dc}{dx} \, dx, \qquad \overline{c\Delta P} = \int_0^L c \frac{dp}{dx} \, dx,$$

and the membrane permeablity and reflection coefficients are defined by

$$\omega = \frac{d^2\alpha^2}{18\eta L \bar{V}_s} (1 - \alpha)\bar{F}, \tag{71}$$

$$\sigma = 1 - \frac{3}{2}(1 - \sigma)\bar{G} \, , \tag{72}$$

where $\bar{F}$ and $\bar{G}$ are average values of F and G across the channel

$$\bar{F} = \frac{2}{1 - \alpha} \int_{a/d}^{1/2} F(\alpha, s)ds \, , \quad \bar{G} = \frac{2}{1 - \alpha} \int_{a/d}^{1/2} G(\alpha, s)ds \, . \tag{73}$$

For dilute solutions J_v, the total volume flux, is closely approximated by the channel water flux $\bar{V}_w J_w$ and the term involving the integral $\overline{c\Delta P}$ can be neglected since $\overline{c\Delta P}$ is of order $\bar{c} \Delta P$ and $\bar{c} \bar{V}_s \ll 1$. Thus equation (69) for a dilute solution reduces to equation (60) where ω and σ are given by (71) and (72). By a similar procedure given in Levitt[37] one can show that L_p is simply the hydraulic resistance coefficient for a constant height Poiseuille channel flow $\frac{d^3}{12\eta L}$. Equation (71) shows that the diffusive permeability ω is related to the average resistance that a freely rotating sphere encounters to parallel translational motion when there is no volume flux. Equation (72) indicates that the reflection coefficient is related to the average slip of a freely rotating particle relative to the Poiseuille profile of the fluid phase.

The expressions for F and G in equations (65) and (66) are
readily computed using the three-dimensional boundary collocation
techniques described in Section 6 when $b/a \geq 1.1$. At closer
particle-to-wall spacings, the full collocation procedure becomes
too time consuming because many boundary points are required for
good accuracy and the F and G functions are rapidly varying. A
more expedient approach is to use an analytic approximation in
which the lubrication theory expression of Goldman et al[39] for F
for the motion of a sphere parallel to a single plane wall is
modified by a function $c(\alpha)$ to account for the nearly constant
effects of the more distant boundary. The desired approximation
for F is

$$ F = \frac{1}{\frac{1}{2}\ln(\frac{b}{a} - 1) + c(\alpha)} \quad , \quad \frac{b}{a} < 1 \tag{74} $$

The function $c(\alpha)$ in (74), which is a constant for any given
channel, is evaluated by matching the lubrication formula (74)
with the values for F computed using the converged two-wall
collocation solutions of Ganatos et al[28] at $b/a = 1.1$.

A similar procedure is used to determine the limiting behavior
of G as $b/a \to 1$. However, in this case difficulty is encountered
due to the fact that the coefficients F_x^p and T_y^p are undefined for
flow bounded by a single plane wall. Since the Poiseuille profile
is almost linear in the vicinity of the wall, this difficult may
be overcome by using a least squares fit between the Poiseuille
profile and a simple shear profile. Thus the lubrication limit of
G/F is given by

$$ \frac{G}{F} \sim F_x^p \sim \frac{b/a}{1.1} \frac{F_x^s}{(F_x^s)_{1.1}} (\frac{G}{F})_{1.1} \quad , \tag{75} $$

where F_x^s is the force coefficient for simple shear flow past a
sphere adjacent to a single plane wall and is given in Goldman,
Cox and Brenner[39] as a function of b/a. The subscript 1.1 in (75)

indicates that the quantity has been evaluated at b/a = 1.1.

The variation of F and G with particle position s and relative particle size α is shown in figure 12. The solid lines are the collocation theory results of Ganatos, Pfeffer and Weinbaum[28]; the dashed lines are the lubrication theory results obtained from (74) and (75). Evidence of the high accuracy of the lubrication theory is found in the continuity of the slope at b/a = 1.1 where the two solutions are matched.

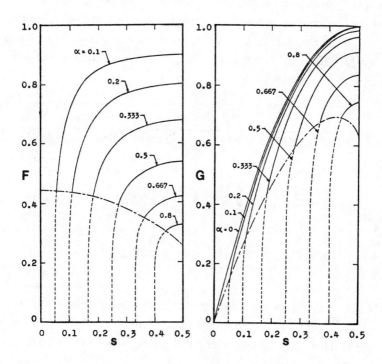

Figure 12.   Variation of F and G functions, equations (65) and (66), with particle position s = b/d for various values of $\alpha = \frac{a}{d}$. ——— three-dimensional strong interaction theory, Ganatos, Pfeffer and Weinbaum[28]; ———— modified lubrication theory, equations (74) and (75).

Calculation of the reflection and permeability coefficients

requires evaluation of the integrals in (73). Accurate numerical evaluation of these integrals over the entire range of s is difficult because of the steep gradients of the F an G functions in the range $a/d \leq s \leq 1.1\ a/d$ (see figure 12). However, in this region, the functions may be evaluated analytically by integrating (74) and (75). These details are given in Ganatos et al[40].

Table 4 shows numerical values of $\omega \bar{V}_s/L_p$ and σ as a function of α. These are compared with the corresponding results obtained by Curry[38] using an ad hoc hydrodynamic model for the slit geometry. Curry's approximate theory substantially underpredicts the value of the reflection coefficient in the range $0 < \alpha < 0.5$ and overpredicts $\omega \bar{V}_s/L_p$ for all values of α the largest erros occuring for $\alpha > 0.5$.

| | σ | | $\omega \bar{V}_s/L_p$ | |
α	Present Study	Curry[38]	Present Study	Curry[38]
0	0	0	0	0
0.1	0.0253	0.015	0.00501	0.0054
0.2	0.0898	0.048	0.0155	0.0170
0.333	0.215	0.142	0.0302	0.0336
0.5	0.413	0.313	0.0407	0.0454
0.667	0.628	0.531	0.0377	0.0428
0.8	0.797	0.719	0.0259	0.0328
1	1	0	0	0

Table 4.   Comparison of the strong interaction theory of Ganatos et al[40] with the approximate results of Curry[38] for the reflection coefficient σ and diffusive permeability ω for the parallel slit geometry.

9.   PARTICLE ENTRANCE EFFECTS; AXISYMMETRIC MOTION.

The flow through finite length pores including entrance and exit effects and the hydrodynamic interaction of particles at the entrance to pores and orifices are classical unsolved problems in low Reynolds number flow. These problems are of

interest in a host of biological applications including unstirred
layer effects in biological and synthetic membranes and filters,
molecular sieving phenomena at the entrance to attached vesicles
(see figure 11) and membrane pores, particle motion near the
mouth of a foraging organism and the Fahraeus effect mentioned in
the introduction. In this section I shall describe how the
strong interaction theory can be extended to obtain exact solutions for the flow through a finite length pore and the axial
motion of a sphere approaching on orifice. In section 10 an
approximate solution will be presented for the three-dimensional
zero drag motion of a sphere approching a pore from any direction.
A more detailed presentation of the theory and results of these
two sections is given in Dagan[41]

Problems involving confining walls which involve more than
one coordinate require special solutions methods since a single
representation of the wall disturbance cannot be written which
is uniformly valid for the entire flow field. The difficulty is
easily illustrated by the two flow geometries shown in figure 13a,
b. Existing solutions to the finite pore geometry shown in sketch
(a) have been limited to (i) the classic solution of Sampson for
the flow through a zero thickness orifice (limiting case of the
hyperboloid coordinate system) and (ii) the flow in an infinite
half space in which the velocity at the pore entrance/exit is
prescribed[42]. The more difficult problem of analytically
determining the interaction between the flow inside and outside
the pore has been examined by several investigators but never
successfully completed. Similarly, the only previous solution
for geometry (b) is the limiting case of a sphere approaching a
solid wall[10]. The difficulty with geometry (b) is that the flow
interaction between the right and left half planes generated by
the motion of the sphere is unknown and must be determined
before the solution in either half space can be completed.

Both of the foregoing problems can be treated by a common
approach in which different streamfunction representations are

168   SHELDON WEINBAUM

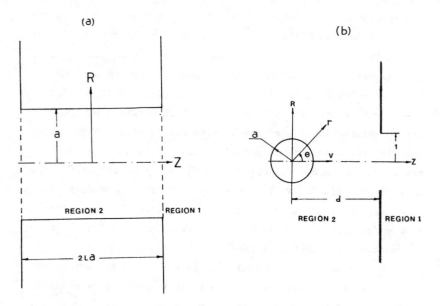

Figure 13.   Flow geometry for (a) a finite length pore and (b) a sphere on the centerline of an orifice.

used in regions 1 and 2 and the two solutions kinematically and dynamically matched in the plane of the orifice or exit. These matching conditions lead to two coupled integral equations to determine the unknown radial and axial velocity components in the matching plane. An outline of this problem formulation is given below.

One starts the solution procedure by prescribing separate kinematic boundary value problems in regions 1 and 2. Thus we require that the no-slip boundary conditions be satisfied on all confining walls in each region

$$\partial\psi_j/\partial R = \partial\psi_j/\partial z = 0, \quad j = 1,2 \tag{76}$$

and that the unknown velocity in the matching plane be prescribed in the form of a Fourier-Bessel series

$$\frac{1}{R}(f(R)\bar{r} + g(R)\bar{z} = \sum_{n=1}^{\infty} [a_n J_1(\lambda_{n,1}R)\bar{r} - b_n(\lambda_{n,0}R)\bar{z}], \qquad (77)$$

where f/R and g/R are the radial and axial velocity components, $\lambda_{n,\nu}$ are the roots of J_ν and a_n and b_n are unknown constant coefficients to be determined.

In region 1, the half plane $z \geq L$ for sketch (a) and $z \geq d$ for sketch (b), the general solution of equation (3) that generates finite velocities at $R = 0$ and as z approaches infinity is

$$\psi_1 = \int_0^{\infty} RJ_1(\alpha R)[A_1(\alpha) + zB_1(\alpha)]e^{-\alpha z}d\alpha , \qquad (78)$$

where $A_1(\alpha)$ and $B_1(\alpha)$ are unknown functions of α and J_1 is the ordinary Bessel function of the first kind.

In region 2 for the finite length pore problem a symmetric solution of (3) about $z = 0$, which generates finite velocities at $R = 0$, is constructed by superposing the disturbances from the cylindrical boundary $R = 1$, the vertical plane $z = 0$ and the unknown flow in the infinite half space $z \geq L$,

$$\psi_2(R,Z) = C_0R^4 + D_0R^2 + \sum_{n=1}^{\infty} [C_nRI_1(\mu_nR) + D_nR^2I_0(\mu_nR)]\cos\mu_nz$$

$$+ \sum_{n=1}^{\infty} [F_nz\sinh\lambda_{n,1}z + G_n\cosh\lambda_{n,1}z]RJ_1(\lambda_{n,1}R) \qquad (79)$$

Here C_n through F_n are unknown constant coefficients, I_0 and I_1 are modified Bessel functions of the first kind, $\lambda_{n,\nu}$ are the roots of J_ν and μ_n are the zeroes of $\sin\mu_nL$. The general solution in region 2 for the sphere and orifice problem is the superposition of a Fourier - Bessel integral equivalent to (78) for the wall disturbance and an axisymmetric spherical disturbance (6)

$$\psi_2(R,Z) = \int_0^{\infty} RJ_1(\alpha R)[A_2(\alpha) + zB_2(\alpha)]e^{-\alpha z}d\alpha$$

$$+ \sum_{n=1}^{\infty} (B_n \rho^{-n+1} + D_n \rho^{-n+3}) I_n(\cos\theta) \qquad (80)$$

The unknown functions $A_1(\alpha)$ and $B_1(\alpha)$ in equation (78) are determined in terms of the unknown velocity at the pore or orifice opening defined by equation (77), by satisfying the no-slip boundary conditions along the exterior wall using Hankel inversion integrals. For the pore problem the streamfunction (79) must also satisfy the velocity distribution (77), if the velocity is to be continuous at the pore opening and satisfy the no slip conditions (76) on the cylindrical surface R=1. Applying these kinematic boundary conditions and utilizing the properties of Fourier-Bessel and Dini series, one can relate the constant coefficients C_n, D_n, F_n, and G_n in expression (79) for the streamfunction inside the pore to the coefficients a_n and b_n for the pore exit velocity. A similar procedure is followed for the sphere-orifice problem except that the unknown functions $A_2(\alpha)$ and $B_2(\alpha)$ in equation (80) are determined by a Hankel inversion of both the spherical disturbance in (80) and the unknown velocity at the orifice opening (77). These Hankel inversion integrals, which also satisfy the no-slip boundary conditions along the wall at z = d, have been evaluated analytically in Dagan[41]. One, therefore, obtains closed form expressions for $A_2(\alpha)$ and $B_2(\alpha)$ in terms of the unknown coefficients B_n, D_n, a_n, and b_n in equations (77) and (80).

For the pore problem the kinematic solution for regions 1 and 2 is now known in terms of the unknown velocity at the pore opening. However, the two solutions have been obtained without accounting for the compatibility of the shear stress and pressure field across the plane of the opening. A unique solution for f(R) and g(R), therefore, requires that the stress tensor be matched at the interface between the two regions. It can be satisfied if the pressure and its gradient are continuous at z = L. The dynamic compatibility conditions for ψ_1 and ψ_2 are therefore,

$$P_1(L^-) = P_2(L^+), \tag{81}$$

$$\frac{\partial P_1(L^-)}{\partial z} = \frac{\partial P_2(L^+)}{\partial z}. \tag{82}$$

To apply (81) and (82) expressions (78) and (79) are substituted in equation (1) and the latter integrated subject to the boundary conditions that $P_1 = P_\infty$ as z approaches infinity and $P_2 = P_o$ at the plane of symmetry. The expressions for P_1 and P_2 are

$$P_1(R,z) = P_\infty + 2\int_0^\infty \alpha J_o(\alpha R)B_1(\alpha)e^{-\alpha z}d\alpha \tag{83}$$

$$P_2(R,z) = P_o + 16C_o z + 2\sum_{n=1}^\infty D_n\mu_n I_o(\mu_n R)\sin\mu_n z$$

$$+ 2\sum_{n=1}^\infty F_n\lambda_{n,1}\sinh(\lambda_{n,1}L)J_o(\lambda_{n,1}R). \tag{84}$$

When the expressions for $B_1'(\alpha)$, C_o, D_n and F_n are substituted into (83) and (84) and the dynamic matching conditions (81) and (82) applied, one obtains a pair of coupled integral equations involving the unknown coefficients a_n and b_n in the velocity profile (77). When the latter equations are integrated over the area of the opening the resulting expressions are recognized as the Fourier - Bessel integral representation of the infinte series (77) for the f and g functions. Evaluating the integral analytically, one obtains a system of linear algebraic equations for the a_n and b_n coefficients involving Struve and generalized hypergeometric functions. The system is then truncated and solved by a standard matrix reduction technique.

Figure 14 shows the steamlines and the pressure distribution for a pore of equal diameter and length. The far field solution was compared with the exact solution[42] for the flow through a zero thickness orifice and found to be in perfect agreement. When $L \geq 1/2$ the flow inside the pore approaches a Poiseuille profile after a short entrance distance of the order of one pore

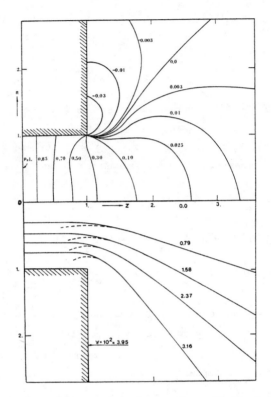

Figure 14.   Steamlines and isobars near a finite length orifice.
Taken from Dagan, Weinbaum and Pfeffer[44].

radius. These results indicate that the entrance effects in a
pore of finite length are of importance only in the vicinity of
the pore opening.

The dynamic matching conditions (81) and (82) also apply for
the sphere-orifice problem with the plane of the opening now
shifted to $z = d$. The expression for the pressure in region 1 is
still given by equation (83), while the expression for the
pressure in the half space containing the sphere is obtained by
integrating equation (1) using the streamfunction representation
(80) for the velocity field. The expression for P_1 is

$$P_1(R,z) = 2 \int_0^\infty \alpha J_0(\alpha R) B_2(\alpha) e^{\alpha z} d\alpha$$

$$+ 2 \sum_{n=2}^\infty D_n \frac{2n-3}{n} \rho^{-n} P_{n-1}(\xi) + P_{-\infty}, \qquad (85)$$

where $P_{-\infty}$ is the pressure at $z = -\infty$.

The matching conditions (81) and (82) are applied in the same
manner as just outlined for the finite length pore problem.
After integrating (81) and (82) across the area of the opening
and solving the set of dual integral equations for the f and g
functions in (77) that result, one obtains solutions for a_n and
b_n that are linear combinations of the spherical coefficients B_n
and D_n. At this point the solutions for ψ_1 and ψ_2 satify the no-
slip boundary conditions on the orifice wall and provide an exact
solution for the velocity at the orifice opening in the form of
an infinite series for an arbitrary coaxial spherical disturbance.
The unknown spherical coefficients B_n and D_n are now determined
by applying the no-slip boundary conditions at discrete points
on the surface of the sphere using the boundary collocation
technique. The truncated matrix equation derived from equation
(80) has the same functional form as equation (30).

The accuracy of the truncated series solution was tested by
a detailed comparison with the exact solution of Brenner[10] for
the drag for the limiting case of a sphere moving perpendicular
to an infinite plane wall in Dagan[41]. Convergence to four
significant digits was achievd for d/a > 1.5 using eight
boundary points and for d/a = 1.1 using 22 equally spaced boundary points.

The application of long standing biological interest is the
sphere-wall entrance effect when a neutrally buoyant particle is
carried by the flow towards the orifice opening. This problem
can be considered as the superposition of two motions, (i) the
axial translation of a sphere with velocity V in a quiescent
fluid and (ii) flow through an orifice past a fixed sphere. Of

particular interest is the ratio of the velocity of the sphere
at any position U(z) to the local centerline velocity of the
fluid U(z) in the absence of the sphere. Equating the drag for
problems (i) and (ii), one can readily show that the condition
for zero drag motion is that

$$\frac{V}{U} = \frac{\lambda^{(u)}}{\lambda^{(v)}} , \tag{86}$$

where $\lambda^{(v)}$ is the drag correction factor for problem (i) and $\lambda^{(u)}$
is the drag correction factor for problem (ii) based on the local
centerline velocity V(z) for the undisturbed flow.

The solutions to equation (86) are plotted in figure 15. The
results indicate that the relative slip between the sphere and
the fluid velocities monotonically increases as the particle
approaches the orifice. The values indicated by arrows on the
ordinate are the ratio of zero drag velocity to Poiseuille
centerline velocity in an infinitely long circular cylinder[12].
Note that the present solutions have only been plotted down to
d/a = 1 (sphere tangent to plane of orifice) since the streamfunction representation (80) is only valid when the sphere lies
entirely in the half space z ≤ d. The solutions for the finite
length pore show that the axial velocity profile in the plane of
the opening lies nearly midway between a fully developed Poiseuille profile and Sampson's solution[42] for a zero thickness
orifice. Maximum local deviations in axial velocity are of the
order of ten percent. One would thus expect that the solution
for V/U for a finite length pore would be nearly the same as for
the zero thickness orifice flow shown in figure 15. The
solutions in figure 15 are an idealized model in the near field of
a food particle entering the mouth of a foraging micro-organism.

10.   THREE DIMENSIONAL MOTION OF A SPHERE APPROACHING A PORE
All the theoretical solutions presented until now were
exact in the sense that if enough terms were retained in the

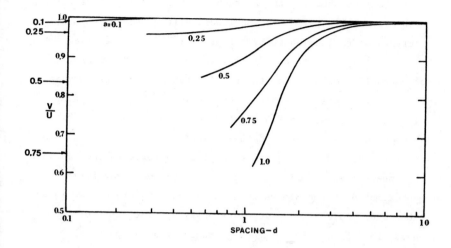

Figure 15.   Ratio of zero drag sphere velocity for neutrally
bouyant particle to local centerline velocity of an undisturbed orifice flow.   a is the ratio of sphere to orifice
diameter.   Taken from Dagan[41].

truncation series for the spheres, convergence to any degree of
accuracy could be achieved since the boundary conditions were
satisfied exactly by analytical methods on both the confining
walls and the matching plane.   For many important three-dimensional flow problems the boundary value problem for the confining
walls is too difficult to treat by exact methods, yet one wishes
to retain the essential features of the strong interaction
analysis.

A well known problem that falls into the above category is the
general three-dimensional motion of a sphere towards a pore of
finite diameter.   This problem is of interest to microcirculatory

physiologists studying the Fahraeus effect and engineers working
with aerosol filtration and nuclepore filters.Recent theoretical
analyses of the nuclepore filter problem[45,46] have focused
primarily on the effects of particle inertia on the impaction
of aerosol droplets near the pore entrance. In these studies
finite difference solutions of the Navier-Stokes equation are
first obtained for the fluid motion through the pore in the
absence of the sphere and this solution then used to calculate
the trajectory of the sphere by applying a force balance in which
the sphere's inertia is equated to the viscous drag force
approximated by the Stokes resistance for a sphere moving in an
unbounded fluid with a local velocity equal to that of the
undisturbed flow. These studies predict that a spherical
particle will enter the pore provided its trajectory does not
come closer to the edge of the orifice than the sphere radius.
All trajectories which lie between the wall and the critical
trajectory are assumed to lead to particle impaction on the walls
of the filter.
        According to nuclepore theory particle trajectories and
fluid streamlines will be identical either in the limit of zero
Reynolds number (all inertia effects vanish) or for a neutrally
buoyant sphere, since the sphere's hydrodynamic interaction with
the orifice wall is neglected. The critical streamline in this
case is defined by the streamline whose closest approach is one
sphere radius from the edge of the pore. The approximate strong
hydrodynamic interaction theory developed by Dagan[41], which I
shall now briefly describe,shows that the critical trajectory
concept is invalid at zero Reynolds number. This theory shows
that a neutrally buoyant particle will always enter the pore
regardless of its initial location and that a particulate
suspension entering a tube from a reservior which is uniform far
from the pore entrance will develop large concentration gradients
near the walls of the orifice as the opening is approached.

        Let us consider a neutrally buoyant sphere of radius a
moving towards a pore of radius c with an instantaneous velocity

(U_x, U_z) at an instantaneous position (X_o, Z_o) in the plane
bisecting the sphere and containing the pore axis. The solution
for the flow through a pore of finite length presented in section
9 has established that the unperturbed flow field is very closely
described by Sampson's solution [42] for the flow through a zero
thickness orifice up to distances of the order of the pore radius
from the edge of the pore opening. Hence, the length of the pore
is immaterial in the formulation of the far field particle
trajectory. Thus, the fluid motion far from the pore entrance
is well defined by Sampson's solution given by

$$\Psi = - \frac{Q}{2\pi}(1 - q^3) \tag{87}$$

where Ψ is the stream function, Q the volumetric flow rate and q
an oblate spheroidal coordinate..

Due to the linearity of the governing equation (1a,b) for
Stokes flow, the general motion of a sphere towards a pore can be
constructed from the superposition of the following asymmetric
flows:

(i)   translation without rotation of the sphere towards the
      pore parallel to the orifice wall, and translation
      without rotation perpendicular to the wall in quiescent
      fluid;

(ii) rotation without translation of the sphere near the
      orifice wall in quiescent fluid;

(iii) flow into the pore past a rigidly held sphere which obeys
      Sampson's solution at infinity.

These three flow situations approximate the three-
dimensional motion of the sphere. The flow in the plane of the
pore entrance is unknown but is not required in the solution procedure, since the matching conditions in this plane will be relaxed.

The force and torque acting on a sphere translating without
rotation parallel to the orifice wall with velocity U_x are defined as follows:

$$\bar{F} = 6\pi\mu a U_x F_x^t \, \bar{i} \qquad \bar{T} = 8\pi\mu a^2 U_x T_y^{t,x} \, \bar{j} \, , \qquad (88)$$

while for the translation perpendicular to the orifice wall with velocity U_z they are given by

$$\bar{F} = 6\pi\mu a U_z F_z^t \, \bar{k}, \qquad \bar{T} = 8\pi\mu a^2 U_z T_y^{t,z} \, \bar{j}. \qquad (89)$$

For a sphere rotating with angular velocity Ω about the y axis in quiescent fluid the force is written in terms of its components as follow:

$$\bar{F} = 6\pi\mu a^2 \Omega F_x^r \, \bar{j} \; + 6\pi\mu a^2 \Omega F_z^r \, \bar{k} \; , \qquad (90)$$

and the torque is

$$\bar{T} = 8\pi\mu a^3 \Omega T_y^r \, \bar{j} \; . \qquad (91)$$

The force and torque acting on a stationary sphere in a Sampson orifice flow at infinity can be treated as the superposition of its two velocity components. Thus, for the radial component of the flow

$$\bar{F} = 6\pi\mu a V_r F_x^s \, \bar{i} \; , \qquad \bar{T} = 4\pi\mu \frac{a^3}{Z_o} V_r T_y^{s,x} \, \bar{j} \, , \qquad (92)$$

where V_r is the undisturbed local radial fluid velocity and the superscript s represents Sampson's solution for the flow through an orifice. For the axial component of the flow

$$\bar{F} = 6\pi\mu a V_z F_z^s \, \bar{k} \; , \qquad \bar{T} = 6\pi\mu \frac{a^3}{X_o} V_z T_y^{s,z} \, \bar{j}, \qquad (93)$$

where V_z is the local undisturbed axial fluid velocity. The undisturbed fluid velocity components V_r and V_z can be obtained directly from the streamfunction solution (87).

The total force and torque acting on the sphere is simply the vector sum of the individual contributions, i.e.

$$\begin{aligned}
\bar{F} = 6\pi\mu a \; & [(U_x F_x^t + V_r F_x^s + a\Omega F_x^r) \, \bar{i} \\
& + (U_z F_z^t + V_z F_z^s + a\Omega F_z^r) \, \bar{k} \,]
\end{aligned} \qquad (94)$$

$$\bar{T} = 8\pi\mu a^2 (U_x T_y^{t,x} + U_z T_y^{t,z} + \frac{a}{2Z_0} V_r T_y^{s,x}$$

$$+ \frac{a}{2X_0} V_z T_y^{s,z} + a\Omega T_y^r)\ \bar{j}. \tag{95}$$

For the special case of interest, a neutrally buoyant sphere being carried by the flow into the pore, the conditions of zero force and torque on the sphere require

$$U_x F_x^t + V_r F_x^s + a\Omega F_{\bar{x}}^r = 0, \tag{96}$$

$$U_z F_z^t + V_z F_z^s + a\Omega F_z^r = 0, \tag{97}$$

$$U_x T_y^{t,x} + U_z T_y^{t,z} + \frac{1}{2Z_0} V_r T_y^{s,x} + \frac{1}{2X_0} V_z T_y^{s,z} + a\Omega T_y^r = 0. \tag{98}$$

Simultaneous solution of these equations for the sphere's translational and angular velocity yields,

$$a\Omega = [T_y^r - \frac{F_x^r}{F_x^t} T_y^{t,x} - \frac{F_z^r}{F_z^t} T_y^{t,z}]^{-1}.$$

$$\{\ V_r[F_x^s/F_x^t\ (T_y^{t,x}) - a/2Z_0(T_y^{s,x})] + V_z[F^s/F^t\ (T_y^{t,z})$$

$$- a/2X_0\ (T_y^{s,z})]\ \} \tag{99}$$

$$U_x = -\frac{1}{F_x^t}\ (V_r F_x^s + a\Omega F_x^r) \tag{100}$$

$$U_z = -\frac{1}{F_z^t}\ (V_z F_z^s + a\Omega F_z^r)\ . \tag{101}$$

From (100) and (101) the governing equation for the trajectory of the sphere is

$$\frac{dZ_0}{dX_0} = \frac{F_x^t}{F_z^t}\ \cdot\ \frac{V_z F_z^s + a\Omega F_z^r}{V_r F_x^s + a\Omega F_x^r}, \tag{102}$$

where V_r and V_z are given by equation (87), Ω by equation (99) and

the right hand side of (102) is evaluated at the center of the
sphere (X_0, Z_0). Clearly, the exact sphere trajectory can be
obtained by integrating (102) provided the hydrodynamic interaction coefficients defined by equations (87) through (93) are unknown.

At present there are no exact solutions for the hydrodynamic
coefficients in equations (99) and (102). To determine these
coefficients one would have to solve each of the flow problems
identified in equations (88)-(93). An equivalent task was described in section 6 where strong interaction solutions were presented for the various force and torque coefficients for a sphere in
parallel channel flow. A simpler approximate procedure is to perform an order of magintude analysis of the relative size of each
of the coefficients to see if equation (102) can not be simplified. This order of magintude analysis is described in detail in
Dagan[41]. It assumes that for small sphere to wall spacings the
transverse curvature effect of the circular pore on the hydrodynamic correction coefficients is much smaller than the resistance
to motion introduced by the planar boundary. Hence, the force
and torque are essentially similar to those acting on a sphere
near an infinite plane wall under equivalent two-dimensional flow
conditions. The final result of this simplication is that equation
(102) is replaced by

$$\frac{dZ_0}{dX_0} = \frac{V_z}{V_r} \left[\frac{F_z^s}{F_z^t} \cdot \frac{F_x^t}{F_x^s} \right] . \qquad (103)$$

The approximate equation (103) excludes the effects of rotation on
the sphere trajectory and the transverse curvature effect of the
circular pore. In order to solve equation (103) it is necessary to
obtain numerical values for the translational force coefficients
F_x^t and F_z^t and the force coefficients F_x^s and F_z^s arising from the
flow past a rigidly held sphere into the pore. The approximate
evaluation of these coefficients is summarized below.

Due to the small induced velocity at the pore opening in

pure translatory motion of the sphere, the motion can be approximated by translation towards an infinite plane wall, provided the sphere is not in the vicinity of the pore opening. Numerical results for the hydrodynamic resistance coefficient for a sphere translating perpendicular to an infinite plane wall are given in Brenner[10] and will be used to represent F_z^t. The force coefficient representing F_x^t for motion parallel to an infinite wall was first obtained by O'Neill[8] and is presented in Goldman et al[11].

The flow past a stationary sphere into a pore can be separated, as already discussed, into flow parallel to the orifice wall and flow perpendicular to the wall where the far field satisfies Sampson's solution equation (87) for the flow through an orifice. When the sphere is rigidly held near the orifice wall the Sampson velocity component V_R parallel to the wall resembles simple Couette flow. Neglecting the small induced velocity at the pore opening and the transverse curvature effect of the pore, one can approximate the behaviour of F_x^s by the force coefficient due to Couette flow past a sphere held near an infinite plane wall. Numerical values of this coefficient were obtained by Goldman et al[47].

The force coefficient F_z^s arising from the problem of the flow perpendicular to the orifice wall past a stationary sphere near the orifice wall can be approximated by considering the axisymmetric uniform flow past a sphere perpendicular to a disk in the limit when the sphere radius is much smaller than the radius of the disk. This approximation is valid near the orifice wall, since the velocity gradient in the radial direction is small in regions where transverse curvature effects can be neglected. The solution for a uniform flow past a sphere in front of a finite disk is closely related to the axisymmetric sphere-orifice solution presented in section 9 since the disk and orifice are orthogonal boundaries of the oblate spheroidal coordinate system. This solution is given in Dagan et al[48].

To test the accuracy of the approximate solutions obtained
by integrating equation (103) and observe the motion of a sphere
in the immediate vicinity of the pore opening where equation
(103) is not valid, an experimental apparatus was built in which
neutrally buoyant polystyrene spheres were released at various
positions relative to a glass plate with a circular pore whose
diameter was 2.25 times that of the sphere. The plate was
mounted horizontally in a low Reynolds number settling tank
filled with a highly viscous transparent fluid and the flow rate
through the pore controlled by a valve located at the bottom of
the settling tank. The experimentally observed and theoretically
predicted sphere trajectories based on equation (103) are
compared in figure 16. Also shown by dashed lines are the
streamline for the undisturbed flow in the absence of the sphere
and the critical streamline denoted by the angle ϕ for which
particle impaction with the pore would occur if the sphere's
hydrodynamic interaction with the walls of the orifice were
neglected.

The results shown in figure 16 are in good agreement with
the predicted theoretical trajectory in the far field and near
the orifice wall. In the vicinity of the pore, the agreement is
good when the initial position is close to the axis of the pore
or its wall. A sphere approaching the pore at an angle of about
45° in the far field departs from the theoretical curve and is
closer to the fluid streamline. This behaviour results from the
transverse curvature effects which are neglected in the
theoretical formulation. Experimental results for $z < a$ are
not presented because parallax prevented accurate measurements in
that region.

The most striking result observed in figure 16 is that the
sphere always enters the pore even in the extreme case where it
moves parallel to the boundary at minimum gap width. According
to the critical trajectory concept all particles originating on
streamlines whose angle ϕ in the far field was less than 56°

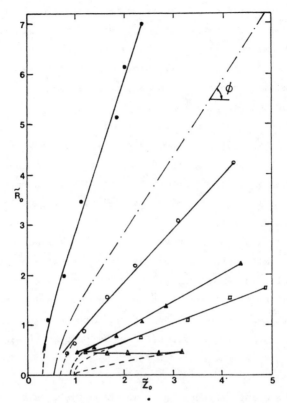

Figure 16.   Comparison between approximate theory and experiment for sphere trajectories. a = 0.5 in., c = 1.125 in.;
——— theoretical solutions equation (103); --- undisturbed
fluid streamlines; ——— · ——— critical trajectory neglecting
wall interaction. Taken from Dagan[41].

would have been excluded from entering the pore for the sphere-
orifice dimensions given. For the case shown this excluded flow
represents approximately 80 percent of the total volumetric flow
through the pore. These results support the more qualitative
predictions of Hocking [49] who argued that the collision
efficiency for small inertialess drops approaches zero when
hydrodynamic interaction effects are taken into account. In

general, collection of particles on the orifice wall will only
occur when adhesive forces at small gap width are significant and
can offset the repelling hydrodynamic forces. In addition, when
particles are small enough for Brownian motion to be important
(less than one micron) the randomizing force due to diffusion
will effect the particle trajectory.

It is evident that when inertial effects are negligible the
undisturbed streamline which passes by the edge of the pore with
a minimum clearance equal to the sphere radius does not properly
describe the exclusion of particles entering the pore. The
hydrodynamic interaction of the particle with the walls of the
orifice, which heretofore has been neglected, will cause large
particles to deviate strongly form fluid streamlines as the
boundary is approached. Since the fluid gap between the sphere
and the wall can never vanish (this would require an infinite
force in creeping motion) the sphere must eventually translate
nearly parallel to the wall and rotate with an angular velocity
approaching $0.5676\dfrac{U_x}{a}$ as demonstrated by Goldman et al[11] for the
free motion of a sphere almost touching a plane wall in a shear
field.

For a dilute suspension in which particles are not collected the entrance effects can be investigated by examining the
steady state average volumetric concentration of particles across
a hemispherical surface of radius r with its origin at the center
of the opening. From physical considerations, when a suspension
of spherical particles flows into the pore, there cannot be any
sphere centers near the orifice wall closer than the sphere
radius. Therefore, one can assume the existence of a particle
depleted wall layer whose thickness is one sphere radius in
which the flux is zero.

Since there is no real exclusion, it is required that

$$H_\infty \bar{V} = H \, \bar{U}_p \qquad\qquad (104)$$

where H_∞ is the concentration far from the pore, $\bar{V}$ is the average

fluid velocity on the hemispherical surface and $\bar{U}_p$ is the average
sphere velocity integrated over the hemispherical surface
excluding the particle free layer. Equation (104) thus defines
some average concentration H of spheres on a hemispherical surface of radius r. To calculate $\bar{U}_p$ the local sphere velocity
components are approximated by equations (100) and (101)
neglecting both the effect of sphere rotation and sphere-sphere
interaction.

The solutions for H are shown in figure 17 for various
particle radii. The average concentration decreases with
decreasing distance r from the pore opening and this concentration defect increases with increasing sphere size. This behaviour is in agreement with the observations made by Fahraeus[1]
showing that when blood flows from a feed reservoir into a tube
of diameter less than 500μm the instantaneous hematocrit within
the tube is always less than that of the blood in the reservoir.

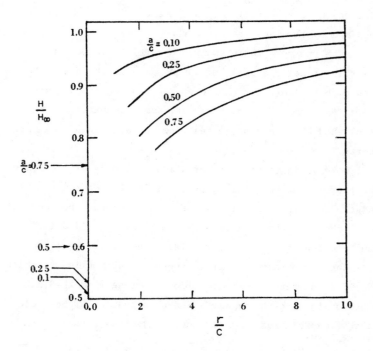

Figure 17.   Average concentration defect H/H_∞ for flow of a
    dilute suspension of spheres into a pore as a function of the
    dimensionless distance r/c along a hemispherical surface with
    origin at the center of the pore opening.

    The results shown in figure 17 were computed for decreasing
values of r until $r = c + 2a$. For smaller values of r the
approximate theory presented in this chapter is not valid because
the effect of the pore on the particle trajectory becomes significant. Nevertheless, the results for small values of r are
compared with the limiting value for a chain of spheres moving
coaxially in an infinite tube, Wang and Skalak[21], shown in
figure 17 by the arrow-heads. Examining these values one can
observe that for large particles ($a/c = 0.75$) the present results
approach this value, while for small particles the concentration
defect is much smaller than the one described by a flow of an
axial chain of spheres in an infinite tube. This behavior can be
explained by the fact that when the particle radius is large compared with the pore radius only one particle can enter the pore
at a time creating a chain of spheres inside the pore. For particles much smaller than the pore, many particles can enter the
pore at the same time and a non-uniform concentration profile is
established which cannot be described by a chain of coaxial
spheres.
    Some insight into the formation of a non-uniform concentra-
tion profile can be obtained by examining the concentration
change along trajectory tubes. For this purpose the generating
arc of the hemispherical surface $\frac{r}{c} = 10$ was divided into nine
equal arc lengths. Since particles cannot cross trajectory
lines one can obtain trajectory tubes by integrating equation
(103) for the particle trajectories starting form the initial
points on the generating arc. The conservation of particle
number flux requires that for each trajectory tube

$$H \int \bar{U} \cdot n \, dS = H_\infty \left(\int \bar{U} \cdot n \, dS \right)_{r/c = 10} = \text{constant}, \quad (105)$$

where the integral is evaluated over the surface area bounded by
the trajectory tube at any distance r. Equation (105) was
evaluated for various sphere radii a and dimensionless distances
$\tilde{r}$ ($\tilde{r} = \frac{r}{c}$), and the local concentration plotted as a function of
θ in figure 18. Clearly, this procedure gives a rough quanti-
tative description of the concentration profile; however, it
provides for the first time a graphic account of the concentration evolution in the flow towards a pore.

Figure 18 shows that the concentration decreases slightly
near the axis of the pore and increases sharply near the orifice
wall. The effect increases with decreasing values of the distance $\tilde{r}$ and the particle radius a. The sharp increase in
particle concentration near the wall as the pore is approached
can be deduced from the sphere trajectories shown in figure 17.
One observes that the sphere trajectory tubes narrow much more
rapidly than the fluid streamtubes near the wall due to their
hydrodynamic interaction with the boundary.

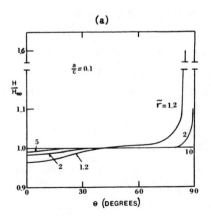
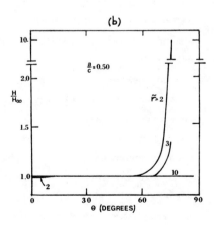

Figure 18.   Sphere concentration as a function of θ in flow
   towards a pore. (a) a/c = 0.1; (b) a/c = 0.5.

   The theory in the present form is unable to predict the
complete transition from the reservoir to tube concentration
because the analysis breaks down in the vicinity of the pore
opening. An important extension of the present theory is to
study this transition region.

11.   CONCLUDING COMMENTS.
            I have tried to convey in this talk some feeling for
both the broad spectrum of heretofore unsolved problems in low
Reynolds number flow that can be solved using boundary collocation, strong interaction methods and the diversity of the
biological problems where this theory can be applied.   Research
is currently underway at the City College to either extend the
analysis or improve existing biological models.   P. Ganatos is
working on the hydrodynamic solutions for the arbitrary motion of
a sphere in a circular cylinder and the tumbling of a ellipsoidal
particle near a planar boundary.   The first problem will be used
to determine the phenomenological coefficients for the cylindrical pore geometry in the Kedem-Katchalsky membrane equations. The
second problem has important application in predicting the
enhanced diffusion of oxygen, plasma proteins and blood platelets
due to the tumbling of reds cells near the endothelial surface of
blood vessels.   As already mentioned, the approximate theory for
the three-dimensional motion of a sphere near a pore entrance is
being improved to include transverse curvature effects near the
opening and the rotation of the sphere.   We hope this will lead
to a more complete understanding of the Fahraeus phenomenon. The
mathematical model for transport along the intercellular cleft
is being generalized to more realistic tight junction geometries
using quasi one-dimensional theory for slowly varying channel
heights. Colleagues at other institutions have started to use the

strong interaction theory to study other particle aggregation
phenomena and the local shearing force on an endothelial cell
arising from the protrusion of its nucleus above the plane of the
vascular interface. The latter problem is of considerable interest to investigators studying the fluid dynamic aspects of
arterial disease. Lastly, I hope this talk will encourage
mathematical biologists with fluid mechanical interests to both
become more familar with the newly developed strong interaction
techniques described herein and to apply them to some of the
numerous other biological problems in their own areas of interest.

BIBLIOGRAPHY

1.   Fahraeus, R., Physiol. Rev. 9 , 241 (1929).

2.   Happel, J. & Brenner, H., Low Reynolds Number Hydro-
       dynamics, 2nd ed. Noordhoff   (1973).

3.   Faxen, H., Arkiv. Mat. Astron. Fys. 17 , No. 27 (1923).

4.   Batchelor, G. K., J.Fluid Mech. 44 , 419 (1970).

5.   Batchelor, G. K., J.Fluid Mech. 52 , 245 (1972).

6.   Oseen, C. W., Ark. F. Mat. Astr. og Fys. 6, No. 29 (1910).

7.   Stimson, M. & Jeffery, G. B., Proc. Roy. Soc. A111, 110
       (1926).

8.   O'Neill, M.E., Mathematika 11 , 67 (1964).

9.   Wacholder, E. & Sather, N.F., J.Fluid Mech. 65, 417 (1974).

10.   Brenner, H., Chem. Eng. Sci. 16, 242 (1961).

11.   Goldman, A.J., Cox, R.G. & Brenner, H., Chem. Eng. Sci.
       22, 637 (1967).

12.   Haberman, W.L. & Sayre, R.M., David W. Taylor Model Basin
       Report No. 1143, Washington D.C. (1958).

13.   Gluckman, N.J., Pfeffer, R. & Weinbaum, S., J.Fluid Mech.
       50, 705 (1971).

14.   Youngren, G.K. & Acrivos, A., J.Fluid Mech. 69, 377 (1975)

15.   Lamb, H., Hydrodynamics, Cambridge University Press,
        Cambridge, (1932).

16.   Ganatos, P.,Pfeffer, R. & Weinbaum, S., J.Fluid Mech. 84,
        79 (1978).

17.   Gluckman, M.J.,Pfeffer, R. & Weinbaum, S., J.Fluid Mech.
        50, 705 (1971).

18.   Davis, A.M.J., O'Neill, M.E., Dorrepaal, .J.M. & Ranger,
        K.B., J.Fluid Mech. 77, 625 (1976).

19.   Landau, L.D., & Lifshitz, E.M., Course of Theoretical
        Physics. In Advanced Physics, vol. 6., Addison-Wesley,
        New York (1959).

20.   Leichtberg, S., Weinbaum, S., Pfeffer, R., & Gluckman,M.J.
        Phil. Trans. Roy. Soc. A 282, 585 (1976).

21.   Wang, H., & Skalak, R., J.Fluid Mech. 38, 75 (1969).

22.   Hyman, W.A. & Skalak, R., Appl. Sci. Res. 26, 27 (1972).

23.   Skalak, R.A., Chen, P.H., & Chien, S., Biorheol. 9, 67
        (1972).

24.   Leichtberg, S., Pfeffer, R., & Weinbaum, S., Int. J.
        Multiphase Flow 3, 147 (1976).

25.   Leichtberg, S., Weinbaum, S., & Pfeffer, R., Biorheol.
        13, 165 (1976).

26.   Ganatos, P., Weinbaum, S., & Pfeffer, R., J.Fluid Mech.
        99, 739 (1980).

27.   Ho, B.P. & Leal, L.G., J.Fluid Mech. 65, 365 (1974).

28.   Ganatos, P., Pfeffer, R., & Weinbaum, S., J.Fluid Mech.
        99, 775 (1980).

29.   Ganatos, P., Weinbaum, S., & Pfeffer, R., "Strong Inter-
        action Solutions for the Gravitational and Zero Drag Mot-
        ion of a Sphere in an Inclined Channel", to be published.

30   Palade, G.E., Anat. Record 136, 245 (1960).

31.   Tomlin, S.G., Biochem. Biophys. Acta. 183, 559 (1969).

32.   Shea, S.M., Karnovsky, M.J., & Bossert, W.H.,   J.Theor.
      Biol. 24, 30 (1969).

33.   Weinbaum, S. & Caro, C.C., J.Fluid Mech. 74, 611 (1976).

34.   Arminski, L., Weinbaum, S., & Pfeffer, R., J.Theor. Biol.
      85, 13 (1980).

35.   Weinbaum, S., and Chien, S., "Vesicular Transport of Macro-
      molecules Across Vascular Endothelium," In Mathematics of
      Microcirculation Phenomena, ed. by J.F. Gross and A. Popel,
      Raven Press, N.Y., 109-132 (1980).

36.   Levitt, D.G., Biophys. J. 15, 533 (1975).

37.   Levitt, D.G., Biophys. J. 15, 553 (1975).

38.   Curry, F.E., Microvascular Res. 8, 236 (1974).

39.   Goldman, A.J., Cox, R.G., & Brenner, H., Chem. Eng. Sci.
      22, 637, (1967).

40.   Ganatos, P., Weinbaum, S., Fischbarg, J., and Liebovitch,L.,
      Advances in Bioengineering, Amer. Soc. Mech. Engineers,
      New York, 193 (1980).

41.   Dagan, Z., Entrance Effects in Stokes Flow through a Pore,
      Ph.D. dissertation, City University of New York, New York
      (1980).

42.   Sampson, R.A., Phil Trans. Roy. Soc. A 182, 449 (1891).

43.   Parmet, I.L., & Saibel, E., Commun. on Pure and Applied
      Math. 19, 17 (1965).

44.   Dagan, Z., Weinbaum, S., & Pfeffer, R., "An infinite Series
      Solution for the Creeping Motion Through Finite Length Pore",
      to be published.

45.   Smith, T.N. & Phillips, C.R., Environ. Sci. Tech. 9, 564
      (1975).

46.   Parker, R.D., & Buzzard, G.H., J.Aerosol Sci. 9 7 (1978).

47.   Goldman, A.J., Cox, R.G., & Brenner, H., Chem. Eng. Sci.
      22, 653 (1967).

48.   Dagan, Z., Pfeffer, R., & Weinbaum, S., "A Strong Inter-
      action Theory for the Axisymmetric Creeping Motion of a
      Sphere Towards Finite Planar Boundaries and Orifices, Part

II: Motion Towards a Disk." to be published.

49.   Hocking, L.M. Q.J. Roy. Met. Soc. $\underline{85}$, 44 (1959).

DEPARTMENT OF MECHANICAL ENGINEERING
CITY COLLEGE OF THE CITY UNIVERSITY OF NEW YORK
140th STREET AND CONVENT AVENUE,
NEW YORK, NEW YORK   10031

Lectures on Mathematics in the Life Sciences
Volume 14, 1981

A MATHEMATICAL MODEL OF HUMAN WALKING

Simon Mochon

ABSTRACT. A series of mathematical models of the swing phase of walking are presented in this paper. In these models, the lower extremities are represented by links and the rest of the body by a point mass. The initial angles and angular velocities of each link are determined by imposing initial (at toeing-off) and final (at heel strike) geometric conditions on the models. The main feature of these models is that they move under the influence of gravity alone, since the total energy of the model is conserved through the swing.

The simplest model possible consists of three links representing the stance leg and the thigh and shank of the swing leg. This model predicts correctly angular excursions of the limbs and the range of possible times of swing. It fails, however, to predict the correct shape of the vertical ground force against time found in normal walking. More complex models are developed by adding progressively all the major determinants of human gait. In this way, the complexity of human walking is divided into its main components and the contribution of each one individual can be evaluated. The discrepancy between the predicted and observed vertical ground force found in the first model is corrected by adding pelvic tilt and ankle plantarflexion to the model. These models can be modified to simulate a range of common gait pathologies.

AMS (MOS) subject classifications (1970).

INTRODUCTION

The analysis of human locomotion began very soon after
the introduction of photography with the work of Marey (1886).
Marey was able to reproduce a whole walking cycle in a single
photographic plate by the use of stroboscopic lamps placed on the
limbs of the subject. In principle, this is all that is required
to get the instantaneous positions, velocities and accelerations
of each of the limbs of the body (kinematic data). Later on,
Braune and Fischer (1889) reported the first data, based on
measurements of cadavers, giving each of the segmental weights as
a ratio of total body weight and the location of the center of
mass of each limb and its segments (anthropometric data). In the
1930s, mostly due to the works of H. Elftman and W. O. Fenn, new
techniques in the analysis of human locomotion were developed and
the understanding of the mechanics of walking and running was
greatly improved. Fenn (1930) built a platform (force plate)
which would record the forces exerted by the foot on the ground
(ground reaction forces). Elftman (1939), using the accelerations of each part of the leg and the ground reaction forces,
calculated forces and moments at the joints of the leg.

These dynamical variables are very important in gait
research and they are not accessible by direct measurements. In
fact, a large part of the research in human locomotion has been
devoted to estimating the muscular forces acting at the joints,
with some studies having the more ambitious aim of predicting the
force produced by each of the muscles of the leg (Hardt (1978)).
In the past 30 years an enormous amount of data has been
gathered describing the kinematics, dynamics and energetics of
walking. Most theoretical approaches have been limited to calculating from experimental kinematic data the variables of gait
which are inaccessible to direct measurement, including the moments at the joints and the energy levels of each of the segments
of the body (Cappozzo et. al. (1975), Thornton-Trump and Daher

(1975), Ralston and Lukin (1969), Cavagna (1966, 1967)). Very few analytical models of walking starting from first principles have been attempted, perhaps because of the complexity of the human musculo-skeletal system.

Among the analytical models, one hypothesis which has been suggested is that walking is performed in a way requiring the least expenditure of energy. In studies examining this hypothesis, optimal trajectories for the limbs are obtained by minimizing some criterion which reflects the mechanical work done by the muscles of the lower extremities. Constant muscle torques are applied to the joints at specific intervals suggested by experimental data and the magnitude of these torques are chosen to minimize the given criterion. In this way, an energy-optimal walking motion is obtained (Beckett and Chang (1968), Chow and Jacobson (1971)). A different approach was presented by Alexander and Goldspink (1977). They used a very simple mathematical model to calculate the power expended during bipedal walking.

Experimental findings in fact suggest that human walking is a learned activity which aims to keep energy consumption as low as possible. To achieve this, the body has to take advantage of the external forces (gravity and ground reactions) to minimize the work that the muscles have to provide. This means, for example, that muscles will be turned off when gravity is acting in the direction of the desired motion. It is, therefore, plausible to expect that whenever possible, the body or parts of it will behave like the coupled links in a compound pendulum. In fact, the suggestion that the action of the swing leg is like the motion of a pure pendulum first appeared in the literature more than a century ago. This suggestion was rejected by many investigators because most swings are performed in times considerably less than the natural half-period of the leg. The purpose of this work is to show that the swing phase of human

gait can be described as a ballistic motion (the swing phase is
the portion of the walking cycle in which only one of the legs is
in contact with the ground. It starts at toe-off and ends at
heel strike). To illustrate this idea, a mathematical model was
developed based on the assumption that no muscular moments are
applied to any of the joints of the leg during the swing period.
Thus the body moves throughout this period under the influence of
gravity alone.

There are few pieces of experimental evidence which make
this assumption plausible. The first, and probably the most important fact about walking, is that the changes in the potential
and kinetic energies of the center of mass of the body are approximately mirror images of one another so that the total energy
is kept at a relatively constant level. This transfer between
kinetic and potential energies is an energy saving process which
reduces the work that the muscles have to apply. In running,
changes in the kinetic and potential energies of the center of
mass are in phase, implying that the mechanical energy of the
body is greatly increased or decreased by muscular activity or
possibly by transient storate of that energy in elastic components. Electromyographic traces also show that the muscles of
the swing leg are reasonably silent during the whole swing
period except just at the beginning and just at the end of the
swing. In accordance with this, the calculated muscular moments
applied at the joints of the swing leg are found to be relatively small.

This paper describes a series of mathematical models of
increasing complexity, which I shall call "ballistic walking
models". One of the aims of this study is to describe human
walking with the least complicated set of ideas. This is valu-
able because it provides a framework for the enormous amount of
existing data. Also, the systematic improvement of the ballistic
walking models to be described divides the complexity of human

walking into its main components and therefore assesses the contribution of each one individually to walking performance.

The models in this paper are organized around the determinants of human gait described by Saunders, Inman and Eberhart
(1953). These are: (1) hip flexion, (2) knee flexion, (3) knee
and ankle interaction of the stance leg, (4) pelvic rotation,
(5) pelvic tilt, and (6) lateral pelvic displacement. Early
versions of the model contain only a few of these determinants;
later versions include progressively more.

BALLISTIC MODELS OF WALKING.

The fundamental assumption of a ballistic walking model
is that once the initial positions and velocities of the legs are
established at the beginning of the swing phase, the model moves
through its stride under the action of gravity alone, and finishes
in a position which allows direct entry into the next step. This
means then that the total energy is conserved.

A model including only the first determinant of gait
(hip flexion) is a system coupling an inverted pendulum representing the stance leg and a simple pendulum representing the
swing leg. It is evident that for this stiff-legged model, the
swing leg will hit the ground at some point during the swing.
To avoid this, flexion of the knee of the swing leg is required
(or walking on a tight rope).

Stiff Stance Leg Model

The first model that will be considered is shown
schematically in Fig. 1. It consists of three links representing
the stance leg and the thigh and shank of the swing leg. The
foot of the swing leg is rigidly attached to the shank. Each
link is assumed to have a distributed mass. The mass of the
trunk, head and arms is represented by a point mass at the hip
joint. The main assumption behind this model is that no muscular moments are applied to any of the joints during the swing

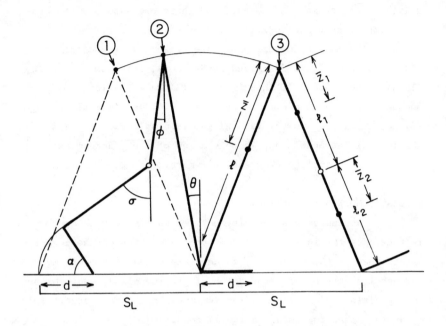

Fig. 1.   Schematic representation of the stiff stance leg model.
The numbers (1), (2), and (3) give respectively the
position of the model at heel strike, toeing off and
following heel strike. The angles, lengths, and positions of the centers of mass of each limb are shown in
the figure. S_L means step length.

period. Therefore, the total energy is conserved and the
equations of motion can be obtained by applying Lagrange's
equations for conservative systems (see Mochon and McMahon
(1980a)).   The equations were transformed to dimensionless form
using the length of the leg and the natural half-period of the
leg as the scales for length and time. The dimensionless constants appearing on the equations were calculated from the
anthropometric data of Dempster. The initial angles and angular
velocities of each link are determined by imposing initial and
final geometric conditions on the model. At toeing off [(2) in
Fig. 1], two conditions specify the position of the toe of the
swing leg on the ground. A third condition fixes the initial
position of the hip by using experimental data of Cavagna (1976)
on the distance that the body moves during the double support
phase, i.e., between (1) and (2) in Fig. 1. At heel strike,
[(3) in Fig. 1], the heel of the swing leg should be in contact
with the ground at a distance S_L from the heel of the other leg;
this imposes two final conditions. Also, the knee angle is required to arrive at zero degrees at the moment of heel strike.
In addition, the constraint that the foot clears the ground must
be satisfied at all times during the swing.

    The results of the computer analysis are shown in Fig.
2. For each step length (S_L) the constraint that the swing leg
must clear the ground sets a lower limit to the possible times
of swing. When the time of swing is made smaller than this
limit (leftmost heavy line) the toe will strike the ground in
mid-swing. Higher values of the time of swing in this model
are found to correspond with increasingly higher maximum values
reached by the knee angle. The next two heavy near-vertical
lines represent swings in which the maximum knee flexion was 90°
and 125°, respectively. Times of swing higher than those given
by the 125° line will be impossible to achieve from the physiological point of view. The 90° line is, however, a more

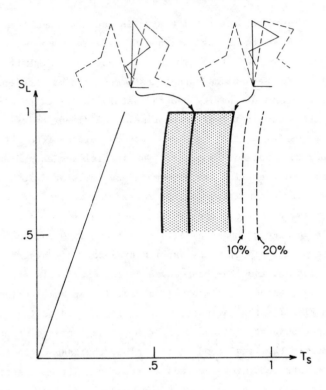

Fig. 2.   The range of possible times of swing for each step
          length for the stiff stance leg model. The upper dia-
          grams show the moment of toe-off, maximum knee flexion
          and maximum hip flexion.

realistic limit, since in normal walking, knee flexion higher
than 90° has not been observed. Note that this model predicts
a range of times of swing between 0.5 and 0.8 times the natural
half-period of the leg, which is within the range of times of
swing found experimentally in normal walking.

Typical angles of the limbs and ground reaction forces
as a function of time predicted by the model during the swing
phase are given in Fig. 3 (forces are normalized with respect to
body weight). Generally, the angles and forces calculated are in
agreement with those found in normal walking. The sole exception
is the vertical force. Experimental traces of the vertical force
during the swing phase show two maxima, one soon after toeing off
and the other at heel strike (both maxima are higher than body
weight). Instead, the predicted vertical force is quite similar
to the one obtained from a simple inverted pendulum (broken lines
in Fig. 3).

In conclusion, we can say that the simple model presented in this section predicts reasonably well the range of
times of swing and angle movements of the limbs. but it fails to
give an accurate description of the vertical force. The main
value of this model lies in the simplicity of the fact that no
empirical data is required to direct its motion; it can be said
that the model "walks by itself". Another important feature of
this model is that it completely disregards the action of
muscles. In these two respects, this model is essentially different from the theoretical studies of walking presented by previous authors, including the studies looking for a minimum
energy criterion as an organizing principle.

Improved Models

The stiff stance leg model includes only the first two
determinants of human gait enumerated in the Introduction. One
possible direction for improving our understanding of human walking is to add progressively to the model the rest of the deter-

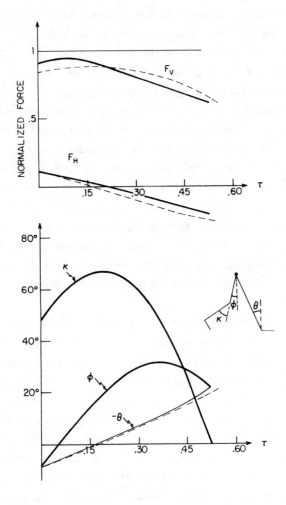

Fig. 3.   Angles and ground reactions predicted by the stiff
          stance leg model (broken lines show the results of a
          simple inverted pendulum).

minants of gait.   In this way we can investigate the effect that
each of them have on walking (see Mochon and McMahon (1980b)).
Here, however, I would like to concentrate on those determinants
that have an effect on the vertical ground force to pin-point
those mechanisms that will improve the lack of agreement found in
the last model between predicted and experimental traces.

One possible mechanism that has been suggested for increasing the vertical force at the end of the swing phase is the
lifting of the heel of the stance leg.   This was investigated
with the model shown in Fig. 4.   It is different from the previous model in two respects: the stance leg has been divided
into two links representing its thigh and shank.   Therefore,
knee flexion of the stance leg has been included in this model.
Secondly, at mid-swing, when the ankle reaches 90° the
model is made to rotate around the middle of the foot, keeping
the ankle locked at 90° until heel strike.   This has the effect
of lifting the heel of the stance leg from the ground.   It is
evident that something is needed to stop the knee of the stance
leg from flexing to large angles.   This, of course, is done in
normal walking by applying muscular moments at this joint.   If a
prescribed moment were to be applied to the knee joint of our
mathematical model, this would violate the ballistic assumption,
since the total energy would no longer be conserved.   This
problem is solved by assuming that the thigh angle θ and the
shank angle α are functionally related: $\theta = \theta(\alpha)$.   This
constraint is equivalent to prescribing the trajectory of the
hip of the model, and, therefore, has the same effect as would
muscular moments of not allowing the knee of the stance leg to
collapse at mid-swing.   This constraint also has the desired
property of keeping the energy conserved (much in the same way
that a bead moving along a frictionless wire under the action of
gravity conserves its mechanical energy.)

The equations of motion, therefore, can be obtained by

204 SIMON MOCHON

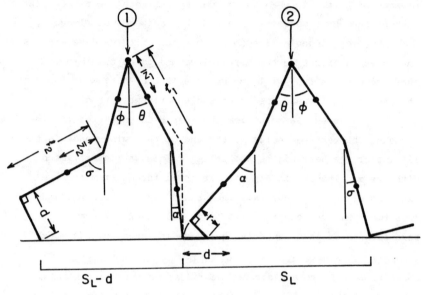

Fig. 4. Model including lifting of the heel and knee flexion of the stance leg.

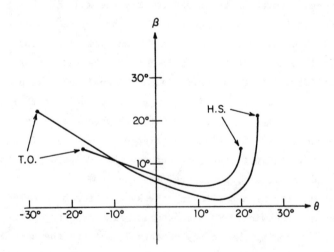

Fig. 5. Curves for two step lengths showing the relations between knee angle $\beta = \alpha - \theta$ and thigh angle θ.

using Lagrange's equations for conservative systems. Since θ should be considered a function of α, the equations of motion contain the functions θ and its derivatives θ' and θ'' (see Mochon and McMahon (1980b) for details).

The relationship between θ and α was taken from experimental data derived from observations of normal walking. Fig. 5 shows two graphs for different step lengths of the function used to describe this relationship ($\beta = \alpha - \theta$). The feature of normal walking that this function tries to reproduce in the model is that the knee of the stance leg is flexed to higher angles as the step length is increased (for small step lengths the knee remains almost fully extended). Using this experimental data and imposing the same geometric conditions and constraint as in the previous model, the equations of motion can be solved. Although the results of this model show some variations from the earlier results, for example, the prediction of a wider range of times of swing, no significant changes are found in predictions for the vertical force. What is more, some swings have a vertical force which goes to negative values at the end of the swing phase. This means that the model will fly off the ground.

Although the predictions of this model might seem disappointing, they in fact show that the lifting of the heel of the stance leg is not the mechanism responsible for increasing the vertical force at the end of the swing phase. The model described in Fig. 4 must be modified still further before the measured and computed vertical force trajectories agree.

This last model, which finally includes movement of the pelvis and the ankle is shown in Fig. 6. A vertical link of length $2p$ joining the hips has been added. Pelvic tilt is then represented by changing the length of this link during the swing. (In general, the pelvic movement also has a horizontal component called pelvic rotation; however, it can be shown that the only effect of pelvic rotation is to change the effective step length).

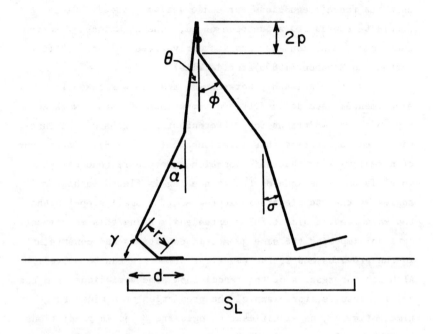

Fig. 6. A complete ballistic model of walking. This model in-
         cludes a vertical pelvic link to simulate the movement
         of the pelvis (pelvic tilt). In addition, the ankle
         angle γ now is allowed to move instead of being fixed
         at 90°, as in the model of Fig. 4.

Also, the ankle angle γ is now specified as a function of α
in order to study the effect of ankle dorsi-plantarflexion.
Again, because θ, p, and γ are given as functions of α, the
ballistic assumption (no change in energy during the swing) holds
true. The general shape of these relations (θ(α), p(α) and
γ(α)) were arranged to conform with the experimental data of
Lamoreux (1971) and Murray (1967).

The results obtained for the vertical force are shown in
Fig. 7. In this figure the corresponding prescribed motions of
the ankle angle and the pelvic link are also given. Comparing
the vertical force predicted by this model with experimental
traces, we can see that they both have the same general shape.
The important features of Fig. 7 are the local peaks in force
appearing just after toeing off and again at heel strike. It is
interesting to note that although both of these movements are
very small (pelvic tilt changes only around 4° and ankle angle
changes by 8°), they have a big effect in the shape of the vertical force. To show more clearly the effect of each of these two
mechanisms, we prescribed to the model different movements of the
ankle joint and the pelvis. The results are shown in Figs. 8 and
9. In Fig. 8, with no pelvic tilt, three vertical force traces
are shown with the corresponding ankle angle movements imposed
in the model. (The dotted line corresponds to an ankle locked
at 90° i.e. ankle angle = 0). We can see from this figure that
ankle angle movement is the mechanism responsible for the second
maximum appearing in the vertical force just before heel strike.
Note that when no movement of the ankle angle is prescribed, the
vertical force goes to negative values and the model flies off
the ground. This suggests that the transition from walking to
running might be due to the inability of the muscles of the ankle
to produce a rapid enough ankle movement to bring the vertical
force above body weight. At this point, walking can no longer
be accomplished and the transition to running occurs. Fig. 9

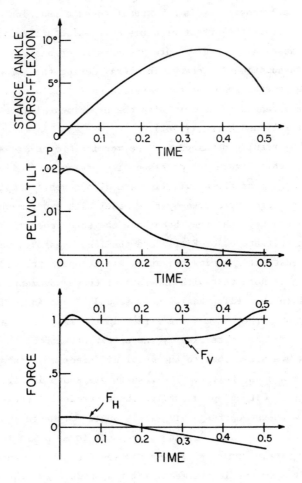

Fig. 7. Ground reaction forces with the prescribed ankle angular
movement and pelvic tilt for the complete model.

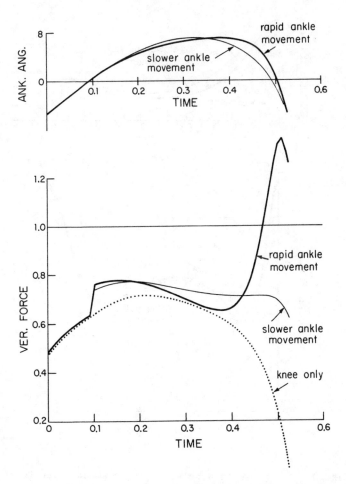

Fig. 8. Two prescribed ankle angle movements with the corresponding vertical ground forces. The dotted line corresponds to an ankle fixed at 90°. No pelvic tilt.

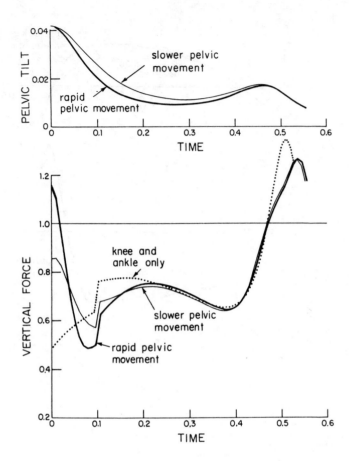

Fig. 9.   Two prescribed pelvic tilt movements with the corresponding vertical ground forces.   The dotted line corresponds to no pelvic tilt.

gives three vertical force traces with the corresponding movement imposed upon the pelvis. (The dotted line corresponds to
zero pelvic tilt). In all these three swings, the ankle angle
movement was the same. This figure shows that the mechanism
responsible for the first maximum immediately after toeing off
is the pelvic tilt movement.

From the results of these three models, it is reasonable
then to claim that many aspects of human walking are well
represented, during the swing phase, by a ballistic model which
moves under the action of gravity alone, thereby, keeping the
total energy at a constant level. These models are offered in
the hope of saying something simple and yet useful about walking,
which is, after all, a very complicated matter.

DIRECTIONS FOR FUTURE RESEARCH

A suggestion for further work concerns better modelling
of the knee. The ballistic walking models presented here predicted correctly the angular excursion of the knee joint during
the swing phase (see Fig. 3). However, at the end of the swing,
the angular velocity of the knee is not zero in these models so
that the shank will bang into the thigh. Of course, this does
not happen in normal walking since the muscles act again immediately before heel strike to stop the knee for a more "smooth"
heel strike. Therefore, to make a more realistic model, a
flexor moment at the knee of the swing leg could be applied at
the end of the swing phase. The magnitude of this moment will be
chosen to satisfy the condition that the angular velocity of the
knee at heel strike should be zero. This extension of the
models is outside the ballistic walking hypothesis since the
moment introduced at the knee can do work on the system and
change its total energy.

Another type of study can be focused on the parameters
of gait and their relationships. A person can walk at a given

speed with a range of possible step frequencies. Nevertheless,
for each speed, there is a preferred step frequency that the individual usually will choose. This preferred relationship between step frequency and walking speed has been correlated with
a minimum of the energy consumption of the individual. Relationships between other parameters of gait, including the time of
swing, velocity, frequency, step length, etc., have also been
reported (Grieve and Gear (1966)). It is interesting to note
that the time of swing varies inversely with speed in adults,
but children show a direct relation between these two parameters
which then changes gradually. This suggests that these relationships are acquired by learning, as the individual selects during
this period the most favorable combination. In general, then,
it is found that parameters of gait are limited to some range
(as in Fig. 2) and that within these ranges, individuals choose
a specific relationship between them. No theoretical work
attempting to predict the form of these relationships has yet
been reported. An optimal type approach like those described in
the introduction could be used for this purpose.

    The models presented in this paper can also be used to
predict the changes in gait that will occur in special situations
like walking on ice, in which the forward force on the ground
can not be allowed to exceed a certain limit, or walking with
heavy shoes, which will be represented by a suitable change of
the constants in the equations of motion.

    A more practically useful application of these models is
the simulation of a range of pathological gaits by introducing
changes into the constraints which are characteristic of various
known pathological conditions. Assessment of these gaits can be
made by determining how much the range of the parameters has been
affected. For example, with muscular weakness of the dorsiflex-
ors of the ankle, the toes hang downward during the swing phase.
This can be simulated by fixing the angle of the ankle of the

swing leg in several degrees of plantar flexion during the swing. Calculation then proceeds as outlined in this paper. The goal is a knowledge of just how severly a given joint immobility or muscular weakness affects walking performance.

The hope is to use the mathematical models presented here to establish a clear delineation between those pathologies of a particular muscle group or joint which are extremely disruptive to normal walking as opposed to those which are not. It may even prove possible to use these models to predict the outcome of a given surgical intervention, so that both the surgeon and the patient could have an improved appreciation of the benefits and risks of a given procedure before it was undertaken.

ACKNOWLEDGEMENTS

I would like to thank Professor T.A. McMahon of Harvard University for his help throughout this study.

REFERENCES

Alexander, R. McN. (1977) Terrestral locomotion. In: Mechanics and energetics of animal locomotion, (Edited by R. McN. Alexander and G. Goldspink), 168-203. Halsted Press, N.Y.

Beckett, R., and Chang, K. (1968) An evaluation of the kinematics of gait by minimum energy. J. Biomech. 1, 147-159.

Cappozzo, A., Leo, T., and Pedotti, A. (1975) A general computing method for the analysis of human locomotion. J. Biomech. 8, 307-320.

Cavagna, G. A., and Margaria, R. (1966) Mechanics of walking. J. Appl. Physiol. 21, 271-278.

Cavagna, G. A., Thys, H., and Zamboni, A. (1976) The sources of external work in level walking and running. J. Physiol. 262, 639-657.

Chow, C. K., and Jacobson, D. H. (1971) Studies of human locomotion via optimal programming. Math. Biosci. 10, 239-306.

Elftman, H. (1939) Forces and energy changes in the leg during walking. Am. J. Physiol. 125, 339-356.

Fenn, W. O. (1930) Work against gravity and work due to velocity changes in running. Am. J. Physiol. 93, 433-462.

Grieve, D. W., and Gear, R. J. (1966) The relationship between the length of stride, step frequency, time of swing and speed of walking for children and adults. Erognomics 9, 379-399.

Hardt, D. E. (1978) A minimum energy solution for muscle force control during walking. Ph.D. Thesis, Massachusetts Institute of Technology.

Lamoreux, L. W. (1971) Kinematic measurements in the study of human walking. Bull. Prosthetics Res. BPR 10-15, 3-84.

Marey, E. J. (1886) La machine animal, (Edited by F. Alcan), Paris.

Mochon, S. and McMahon, T. A. (1980a) Ballistic walking. J. Biomech. 13, 49-57.

Mochon, S. and McMahon, T. A. (1980b) Ballistic walking: an improved model. Math. Biosci. 52, 241-260.

Murray, M. P. (1967) Gait as a total pattern of movement. Am. J. Phys. Med. 46, 290-333.

Ralston, H. J. and Lukin, L. (1969) Energy levels of human body segments during level walking. Ergonomics 12, 39-46.

Saunders, J. B., Inman, V. T., and Eberhart, H. D. (1953) The major determinants in normal and pathological gait. J. Bone Jt. Surg. 35-A, 543-558.

Thornton-Trump, A. B., and Daher, R. (1975) The prediction of reaction forces from gait data. J. Biomech. 8, 173-178.

DEPARTMENT OF MATHEMATICS

MASSACHUSETTS INSTITUTE OF TECHNOLOGY

CAMBRIDGE, MA.   02139

ABCDEFGHIJ–AMS–8987654321